# QUESTIONS AND ANSWERS IN ENVIRONMENTAL SCIENCE PRACTICAL



By

**Rahul K Kamble**

Assistant Professor
Department of Environmental Science
Sardar Patel College, Chandrapur
Maharashtra

Competition Tutor

An imprint of Scientific Publishers (India)

*Published by:*
**Competition Tutor**
5-B, New Pali Road, P.O. Box 91
Jodhpur 342 001 (India)

E-mail: competitiontutor@gmail.com
Website: www.competitiontutor.co.in



An imprint of Scientific Publishers (India)



ISBN: 978-93-87307-33-9

Printed in India

# Preface

Environmental Science comprises of different matrixes such as air, water, soil and living components. To have in-depth understanding of these, practical knowledge becomes very essential. In this regard, practical is one of the important aspects of the subject. These practical's gives an insight to various phenomenons occurring in the environment. The analysis of various environmental matrixes can be understood thoroughly through methodological insight understanding of them. In this regard, an attempt has been carried out through this book, perhaps may be first of its kind, to provide in detail various aspects involved in environmental analysis of air, water, soil and noise.

This book is primarily prepared to cater students of undergraduate, postgraduate, research scholars and faculty members in Environmental Science, Environmental Engineering, Environmental Technology of universities/ institutes of India and abroad. It provides sufficient theoretical and practical knowledge about various environmental parameters, so as to have a clear understanding of them.

The book comprises of four parts viz. air, water, soil and noise. Each part further containes various parameters involved in them except noise. Number of questions and answers on each parameter are presented in lucid and concise manner, so as to make all the aspects of it understandable. In addition to this, a number of appendixes are also upended which will provide additional knowledge on these parameters for overall understanding of them.

Further suggestions from readers for improvement of the book will be highly appreciated on *rahulkk41279@yahoo.com.*

**– Rahul K Kamble**

# Acknowledgments

The completion of this book had received valuable contributions from my students of post graduate, Department of Environmental Science namely, Priya Kolhe, Pratiksha Thamke, Naznin Kazi, Mayur Mehta, Prashik Kamble and others. I am indebted to them. I would like to record my sincere thanks to Ms Namrata R. Nimgade, for her comments on the manuscript and suggestions for its improvement.

The book could not have seen the light of the day without the constant encouragement and support from Mrs Pradnya Kamble– my wife, who had always inspired and motivated me to complete this task. Sincere thanks are also due to Mr Sunil Admane, for typing the manuscript and carrying out corrections into it.

I am humbly grateful to Mr Tanay Sharma, Managing Director, Scientific Publisher, Jodhpur and all other supporting staff members of the publication house for giving me an opportunity in bringing out this book in a short period of time.

**Rahul K. Kamble**

**In Loving Memory of**

**Late Dr. Elizabeth M. Van de Ven**

Former Business Director of

MBA Environmental Management Programme

Cartesius Institute

University of Twente, the Netherlands

# Contents

## Common terminologies used in Environmental Analysis

**Distilled water**: The distilled water is water that has been purified by the distillation process and is free of dissolved salts.

**Deionized water (DI water)**: The deionized water is water that has been prepared by passing it through two resins which attracts positive and negative ions and thus removes all ions from it. This water is also called as DM water or demineralized water.

**$CO_2$ free distilled water**: It is distilled water, boiled for sufficient duration of time so as to remove dissolved $CO_2$ from it. It is then covered and cooled down and used for preparation of reagents.

**Laboratory grade water (Grade I, II, III and IV)**: American Society for Testing and Material (ASTM) had defined four different standard grades for pure water to be used for laboratory analysis. The water quality for these different grade waters is based on conductivity, resistivity, TOC, total silica, sodium, chloride, pH, bacteria and endotoxin content. These waters are also called as Ultrapure Water (UPW).

**Millipore water**: It is an ultrapure water or Type 1 water, prepared by Milli-Q water filtration system.

**Soft water**: Soft water is water from which calcium and magnesium ions have been removed.

**Reagents**: A laboratory chemical used to carry out chemical reactions or formation of coloured compounds.

**AR grade**: Analytical reagent grade (AR grade) contains chemicals used for analytical purposes and research work. These chemicals are of high purity in nature. The chemicals are provided with a certification for minimum assay and maximum limit of trace impurities.

**LR grade**: Laboratory reagent grade (LR grade) contains pure/extra pure chemicals for laboratory analysis.

**Extra pure**: Extra pure are chemicals preferably used in pharmaceutical analysis.

**Atomic weight**: Atomic weight of the elements refers to the relative weights of the atoms as compared with some standards. (The $^{12}C$ isotope of carbon was adopted as the atomic weight standard, with a value of exactly 12). In general, elements do not have atomic weights that are whole numbers because they consist of a mixture of isotopes.

**Gram Atomic Weight**: The gram atomic weight (GAW) of an element refers to a quantity of the element in grams corresponding to the atomic weight.

**Gram Molecular Weight/Mole**: The term gram molecular weight (GMW, the short hand symbol being MW) refers to the molecular weight in grams of any particular compound. It is also referred to as mole.

**Equivalent weight**: Equivalent weight can be defined as follows: EW = MW/Z. Where, Z = the number of $H^+$ or $OH^-$ ions a species can react with or yield in an acid base reaction. One equivalent is defined as one mole of a compound divided by it's EW.

**Valency**: Valency or oxidation state is the same and defines the combining capacity of an element.

**Normality (*N*)**: It is the number of gram equivalent weight of solute dissolved in 1000 mL of solution or number of gram millequivalent of solute per millilitre of solution. It is denoted by *N*. For calculation of normality:

Normality (*N*) = weight/L/ equivalent weight

Example: To prepare 1 *N* $K_2Cr_2O_7$ solution, we will requre

1 *N* $K_2Cr_2O_7$ = weight/L/ equivalent weight

Equivalent weight of $K_2Cr_2O_7$ is 49, then

1 *N* = weight/L/49

Hence, Weight/L= 49 x 1. Thus, 49 gram of $K_2Cr_2O_7$ has to be dissolved in 1000 mL of distilled water to get a solution of 1*N* $K_2Cr_2O_7$.

**Parts per million (ppm)**: It is a weight to weight ratio. A litre of water weights approximately 1000 gram or 10,00,000 mg and hence 1 mg/L is considered to be equivalley 1 ppm.

**mg/L**: It is a weight-volume relationship and, when dealing with liquids, it offers a convenient basis for calculation.

mg/L x 8.34 = lb/million gal, mg/L = g/m$^3$, mg/L x 10$^{-3}$ = kg/m$^3$.

**Molar solution (M)**: A molar solution consists of 1 gram molecular weight dissolved in enough water to make 1000 mL of solution. The amount of molecular weight in grams dissolved in 1000 mL of solution.

**Molal solution**: One gram molecular weight dissolved in 1000 mL of water, then resulting solution having a volume slightly in excess of 1000 mL.

**Primary standard**: Primary standard is a solution for which a solution of a definite concentration can be prepared. It is stable in nature. Example: Potassium dichromate, Sodium chloride etc.

**Secondary standard**: Secondary standard is a solution which is prepared by standardizing against primary standard solution so as to prepare the exact concentration of it. It is unstable and reactive in nature. Example: Sodium hydroxide.

**Indicator**: A solution/powder used during titration process so as to know the end point of the titration process.

**Titration**: Titration is a process in which a titrant is added from a burette into a sample usually liquid (water/waste water) into a conical flask so as to complete the reaction between these two solutions.

**Titrant**: A solution of definite concentration (strength) filled into a burette and used for titration.

**Titrate**: A solution taken into a conical flask for titration.

**End point**: End point is a point in a titration process in which titrant reacts with titrate so as to be chemically equivalent. The end point is denoted by change in colour of the titration.

**Aliquots**: Aliquots are serial dilutions prepared from a known strength of solution but having much lower concentration than it or the amount of a sample used for analysis.

**Calibration curve:** In analytical chemistry, a calibration curve, also known as a standard curve, is a general method for determining the concentration of a substance in an unknown sample by comparing the unknown to a set of standard samples of known concentration.

**Standard method**: This is an abbreviation for the book entitled Standard Methods for the Examination of Water and Wastewater published jointly by the American Public Health Association, American Water Works Association, and Water Environment Federation. This book contains various analytical methods for water and wastewater analysis. This book is the primary reference book and new edition is published after every five years. In the year 1905, the First edition was published and as of 2017, 22nd edition is available.

**Ambient environment**: Outside existing natural environment.

# 1

# AIR

## ■ 1.1 Suspended and Respirable Particulate Matter

**Q 1. What is a high volume air sampler?**

- High volume air sampler (HVS) is a type of instrument that is used to draw atmospheric air through a filter paper to measure suspended, respirable and fine particulate matter as well as different gaseous pollutants.

**Q 2. State components of high volume air sampler.**

- Components of high volume air sampler includes:

i. The face plate and gasket
ii. The filter adapter assembly and
iii. The motor or blower unit.

**Q 3. Describe the filter paper used in HVS.**

- The filter paper used in HVS is of borosilicate glass fibre filter paper of size 8×10 inches, with an exposed area of 400 $cm^2$ and pore size of 1-3 micron.

**Q 4. How ambient air sampling site is identified?**

- For identification of ambient air sampling site, the following points needs to be considered:
  - i. Sampler should have free exposure so that sample is collected by setting only
  - ii. It should not have any interference from buildings, high rise objects etc., and
  - iii. No stacks or chimneys should be present in the vicinity of the site.

**Q 5. What is the sampling duration for HVS?**

- Sample for HVS should be carried out continuously for 24 hours. However, if SPM concentration is higher, duration may alter to 6, 8 or 12 hours. This can be decided after field survey.

**Q 6. What types of pollutants can be sampled through HVS?**

- SPM, RSPM, fine particulate matter and various gaseous pollutants are sampled through HVS.

**Q 7. What is the alternative name for personal sampler?**

- Low volume sampler is the alternative name for personal sampler.

**Q 8. Where low volume air sampler is used?**

- Low volume air sampler is used to determine indoor air quality.

**Q 9. Where HVS is used?**

- HVS is used for ambient air sampling to determine ambient air quality.

**Q 10. Why monitoring of indoor air pollutants is important?**

- Monitoring of indoor air pollutants is important because human being spends about 80% of their time in indoor environment and thus it is important to identify various adverse health effects by them.

**Q 11. What is the use of sampled particulate matter on glass fibre filter paper?**

- Sampled particulate matter on glass fibre filter paper can be used for heavy metal analysis, factor analysis, hydrocarbon analysis and microbial analysis.

**Q 12. Why is the study of air pollutants important?**

- Study of air pollutants is important as these pollutants cause adverse effects on human health and surrounding environment.

**Q 13. Why SPM is not sampled by dust fall jar method?**

- Particles in the size range of 1 to 100 micron (SPM) cannot be sampled by dust fall jar method because of non-settleability of them by gravity.

**Q 14. SPM determination is based upon which method.**

- SPM determination is based upon gravimetric method.

**Q 15. State various methods for sampling air pollutants.**

- HVS, dust fall jar, sulphation rate are some of the methods for sampling air pollutants.

**Q 16. What is the sampling frequency of monitoring through HVS?**

- HVS sampling should be carried out for 2 days per week and 104 sampling days per year.

**Q 17. Name the apparatus used for sampling of gases in HVS.**

- Greenburg Smith impingers are used for sampling (absorbing) gaseous pollutants in HVS.

**Q 18. State the types of impingers.**

- There are two types of impingers; Dry impinger and Wet impinger.

**Q 19. How is ambient air sample collected in an impinger?**

- Ambient air sample is drawn with the help of HVS in an impinger through an orifice of 2.3 mm in diameter and against a flat surface of 5 mm distance, with a jet velocity of about 113 m/s at a sampling rate of 28.3 L/min.

**Q 20. State the types of wet collector.**

- Greenburg Smith standard impinger and Midget impinger are the types of wet collector.

**Q 21. Describe Midget type of collector.**

- In Midget type of collector air is drawn at a rate of 2.8 L/min through an orifice of 1 mm in diameter and impinger against a flat surface of 5 mm distance at a jet velocity of about 60 m/s. The collection efficiency is low for particles less than 0.74 micron in diameter.

**Q 22. HVS is based upon which law?**

- HVS is based on Stokes law.

**Q 23. What is the disadvantage in using glass fibre filter paper?**

- The significant disadvantage in using glass fibre filter paper is its tendency to absorb atmospheric humidity.

**Q 24. State several forms of organic membrane filters used for particle collection.**

- The most common forms of organic membrane filters used for particle collection includes Millipore, Nuclepore, Pallflex and Aquapore which are composed of cellulose acetate or nitrocellulose and fluoropore miller which are teflon filters.

**Q 25. Why is glass fibre filter paper mostly used in air pollutant sampling?**

- Glass fibre filter paper is inert to acid and to organic solvents and thus is mostly used for air pollutant sampling.

**Q 26. How is the volume of air sampled in HVS calculated?**

- The volume of air sampled in HVS is calculated with the help of the formula: Volume of air sampled ($m^3$) = Average manometer reading × Sampling duration (min.).

**Q 27. In which rang, SPM is obtained?**

- SPM is obtained in the range of 10 to 100 μ detanid in diameter.

**Q 28. What is the flow rate in HVS for particulate matter sampling?**

- The flow rate in HVS for particulate matter sampling is 1.1 to 1.5 $m^3$/min.

**Q 29. What is the minimum concentration of SPM that can be detected by HVS?**

- The minimum concentration of SPM that can be detected by HVS is 10 $\mu g/m^3$.

**Q 30. What is an another name for HVS?**

- Another name for HVS is Repairable Dust Sampler (RDS).

**Q 31. What is the pore size of glass fibre filter paper?**

- The pore size of glass fibre filter paper is 0.3 μ.

**Q 32. Which trace element impurities are present in glass fibre filter paper?**

- Glass fibre filter paper contains substantial amounts of several trace elements impurities like Cr, Cu, Fe, Mn, Ni, Pb, Sb and Zn.

**Q 33. Why it is necessary to collect large volume of air sample in HVS?**

- Collection of large volume of air sample in HVS is necessary to reduce contribution of impurities.

**Q 34. In which unit is the concentration of air pollutants reported?**

- The concentration of air pollutants is reported in terms of milligrams per cubic meter ($mg/m^3$) or micrograms per cubic meter ($\mu g/m^3$).

**Q 35. How is the flow rate in impinger of HVS monitored/adjusted?**

- Flow rate in impinger of HVS is monitored/adjusted with the help of a rotameter.

**Q 36. In which situation solution impingers are employed.**

- Solution impinger may be employed in a situation where high atmospheric particulate loadings exists.

**Q 37. What is the exposed area of glass fibre filter paper in a HVS?**

- The exposed area of glass fibre filter paper in a HVS is 400 $cm^2$.

**Q 38. How many types of particulate matter are there?**

- Types of particulate matter includes: SPM: Suspended particulate matter (size between 10-100 micron), PM10: Particulate matter 10 or respirable particulate matter (RPM) (size less than 10 micron) and PM2.5: Particulate matter 2.5 or fine particulate matter (size less than 2.5 micron).

**Q 39. How is SPM concentration calculated?**

- SPM concentration is calculated with the formula:

$$\text{SPM}\left(\mu g/m^3\right)=\frac{\text{Difference in weight of glass fibre filter paper }(g)\times 10^6}{\text{Flow rate }\left(\text{litre per minute}\right)\times\text{ total run in minute}}$$

**Q 40. Why is manometer used in HVS?**

- Manometer is used to check flow rate of incoming ambient air through a blower into a HVS.

**Q 41. What is the range of particle matter collected on glass fibre filter paper?**

- Particle matter collected on glass fibre filter paper is in the size range of 0.1 to 100 microns in diameter and further depends on type of HVS used.

**Q 42. What should be the space between the hood and the body of the HVS?**

- The space between the hood and the body of the HVS should be around 600-700 $cm^2$ so that suspended particulate matter can enter into it.

## 1.2 Dust Fall

**Q 1. What are particulates?**

- Small solid particles and liquid droplets are collectively called as particulates.

**Q 2. State sources of particulate matter.**

- Sources of particulate matter includes thick, black smoke bleaching out of exhaust pipes of vehicles, swirls of dust picked up by wind, ash and soot coming from campfires.

**Q 3. What is dust?**

- Dust is small, solid particle created by breakup of larger masses through processing such as crushing, grinding or blasting, that may come directly from processing or handling of materials such as coal, cement or grains.

**Q 4. What is dust made up of?**

- Dust is made up of solid particles predominantly larger than those found in collides and capable of temporary suspension in air or other gases.

**Q 5. How dust produced?**

- Dust is produced by crushing and grinding of organic and inorganic materials.

**Q 6. What is the size of dust?**

- The size of dust is above 20 μ in diameter and some are smaller in size.

**Q 7. What health problems are associated with air pollutants?**

- Health problems associated with air pollutants include breathing problem, asthma, reduced lung function, lung damage, cancer, brain and nervous system damage.

**Q 8. Which activities are responsible for dust formation?**

- Activities responsible for dust formation includes manufacturing, domestic and industrial activities.

**Q 9. Which industries contribute to atmospheric dust?**

- Construction, agriculture and mining are among the industries that contribute most of the atmospheric dust.

**Q 10. Name the activities which contribute to dust generation.**

- Vehicle entrainment of dust from paved and unpaved roads, wind erosion of open areas, stockpiles and tailings impoundments, drilling and blasting operations activities contribute to dust generation.

**Q 11. In which range dust particles are obtained?**

- According to the "Glossary of Atmospheric Chemistry Terms", dust particle ranges in size from 1 to 100 μ in diameter.

**Q 12. How particles less than 1 micron in diameter are classified?**

- Particles less than 1 micron in diameter are classified as smoke or fumes and falls within fine mode range.

**Q 13. What are the impacts of dust on Human beings?**

- Excess concentration of dust in workplace may reduce visibility, cause unpleasant deposits in eyes, ears and nasal passages and may cause injury to skin or mucous membranes by chemical or mechanical action.

**Q 14. What are the sources of dust?**

- Sources of dust includes soil disturbance during construction projects, disturbed land areas that are cleared and vacant, unpaved roads, parking lots and playgrounds, windblown emissions from tiled fields, military training exercise, unpaved equipment yards and undisturbed desert areas during the highest winds.

**Q 15. What is aerosol?**

- Suspended solid or liquid dust particles in atmosphere are termed as aerosol.

**Q 16. What are Primary pollutants?**

- Primary pollutants are those pollutants which are emitted from an identifiable source.

**Q 17. Why wind borne dust is important?**

- Wind borne dust is important as it affects health due to entry into respiratory tract or as a secondary pathway for ingestion of toxic materials.

**Q 18. What are ecological effects of dust?**

- Wet and dry deposition of dust causes damage to plants, metal surfaces, fabrics and buildings.

**Q 19. What are the primary effects of particulate matter on vegetation?**

- Primary effects of particulate matter on vegetation includes reduced growth and productivity due to interference with photosynthesis and phototoxic impacts.

**Q 20. What are the effects of dust on animals?**

- Deposition of acidic dust in aquatic system alters pH and results in acidification of lakes and ponds. Low pH kills aquatic life forms and results into lakes with no fish species.

**Q 21. How should the period of dust sampling be determined?**

- The period of dust sampling should be determined by the process and installations of producing dust.

**Q 22. What are particles less than 1 micron classified as?**

- Particles less than 1 micron are classified as smoke or fumes and fall within fine mode range.

**Q 23. How are dust particles larger than 1 micron formed?**

- Dust particles larger than 1 micron are formed by mechanical disintegration of material such as crushing, grinding and blasting.

**Q 24. On which factors impact of particles on human health depends upon.**

- The impact of particles on human health is largely dependent upon:
  - i. Particle characteristic particularly particle size and chemical composition and
  - ii. The duration, frequency and magnitude of exposure.

**Q 25. Which reagents are used in dust fall jar method?**

- Reagents such as copper sulphate and mercuric chloride are used in dust fall jar method.

**Q 26. Name the sampler used to determine dust fall rate.**

- Dust fall jar (usually a plastic container) is used to determine dust fall rate.

**Q 27. Which filter paper is used to filter dust sample collected in a dust fall jar?**

- Whatman filter paper No. 42 is used to filter dust sample collected in a dust fall jar.

**Q 28. Which balance is used for weighing of dust sample?**

- A balance with sensitivity of 0.1 gram and a range adequate for weighing of both crucible and filter paper are used for weighing dust sample.

**Q 29. Why oven is used in dust fall jar analysis?**

- An oven is used in dust fall jar analysis to evaporate water from a portion of filtered liquid placed in a crucible.

**Q 30. Which sieve is used to remove large particle from a sample?**

- Sieve no. 18 (1 mm mesh size) is used to remove large particles from a sample.

**Q 31. Which anti-freezing agent is used in dust fall jar estimation?**

- Isopropyl alcohol is used in dust fall jar estimation to avoid freezing of water.

**Q 32. What are interferences in dust fall jar sampling?**

- Bird droppings, insects, algae, fungi and large plant material are interferences in dust fall jar sampling.

**Q 33. How can bird droppings be minimized?**

- Bird droppings can be minimized by using a bird ring/guard.

**Q 34. How can particles larger than 1 micron diameter be removed?**

- Particles larger than 1 micron diameter can be removed by sieving the sample through a sieve.

**Q 35. Why soluble and insoluble deposit analysis is being carried out?**

- Soluble and insoluble deposit are subjected to analysis for ions, elements, inorganic and organic compounds.

**Q 36. Describe particle fallout collector used in dust fall jar estimation.**

- A particle fallout collector used in dust fall jar estimation comprise of cylindrical container usually made up of plastic with a water holding capacity of about 3-5 litres located on a stand at least 1.2 m above the support surface.

**Q 37. What should be the sampling period for dust fall rate?**

- The sampling period for dust fall rate should be of 30 days i.e. 1 month.

**Q 38. Why anti-freezing agent is added in water sample of dust fall jar?**

- Anti-freezing agent is added in water sample of dust fall jar to prevent freezing of water that is used for collection of dust sample into the container.

**Q 39. In which places use of anti-freezing agent is necessary?**

- Anti-freezing agent is necessary in places with colder climates or with higher precipitation levels.

**Q 40. Where should be the rim container placed?**

- The rim container should be placed between 2.4 m and 15 m above ground level and at least 1.3 m above the supporting surface.

## 1.3 Sulphur Dioxide

**Q 1. What are the sources of atmospheric sulphur dioxide?**

- Burning of fossil fuels in thermal power plant, manufacturing of sulphuric acid, fertilizers and smelting industries are some sources of atmospheric sulphur dioxide.

**Q 2. Which method is used to estimate atmospheric sulphur dioxide concentration?**

- Modified West and Gaeke method is used to estimate atmospheric sulphur dioxide concentration. In the previous method sulphamic acid was not used; however, in the modifid version, it is used to remove $NO_x$.

**Q 3. Which absorbing media is used to sample atmospheric sulphur dioxide gas?**

- 0.1 *N* Potassium tetrachloro mercurate (KTCM) is used as an absorbing media for sampling atmospheric sulphur dioxide gas.

**Q 4. How KTCM (Potassium tetrachloro mercurate) absorbing media is prepared?**

- KTCM (Potassium tetrachloro mercurate) absorbing media is prepared by dissolving 10.86 g mercuric chloride, 0.066 g EDTA disodium salt and 6 g KCl in 1000 mL double distilled water.

**Q 5. At which wavelength on spectrophotometer, sulphur dioxide concentration is measured?**

- Sulphur dioxide concentration is measured at 560 nm on spectrophotometer.

**Q 6. Which formula is used to estimate sulphur dioxide concentration?**

- Sulphur dioxide concentration is measured with the help of formula:

$$\text{Sulphur dioxide in } \mu g/m^3 = \frac{\text{Conc. of } SO_2 \text{in sample from graph } (\mu g) \times 1000}{\text{Flow rate in HVS (litres per minute)} \times \text{Total run (min.)}}$$

**Q 7. Which complex is formed in sulphur dioxide estimation?**

- A dichlorosulphitomercurate complex is formed in sulphur dioxide estimation.

**Q 8. Which impinger is used to absorb atmospheric sulphur dioxide gas?**

- Midget impinger is used to absorb atmospheric sulphur dioxide gas.

**Q 9. At which flow rate sulphur dioxide is absorbed into the absorbing media.**

- Sulphur dioxide is absorbed into the absorbing media at a flow rate of 0.5 litres per min (LPM).

**Q 10. What are the factors determining efficiency of impingement?**

- Factors determining efficiency of impingement depends upon velocity of impingement, impinging distance and depth of emission.

**Q 11. What are wet impingers?**

- Wet impingers are those in which liquid absorbing media is used.

**Q 12. What are impactors?**

- Dry impingers are also known as impactors.

**Q 13. How atmospheric gases are absorbed in absorbing media?**

- When atmospheric air containing different gases passes through an absorbing media it react with it. As flow rate of incoming gas is very slow (0.5 litres per minute) these gases gets completely dissolved into absorbing media.

**Q 14. What is the significance of atmospheric sulphur dioxide estimation?**

- Estimation of atmospheric sulphur dioxide has following significance:

  i. $SO_2$ is absorbed by nasal system in human being leading to swelling and stimulated muscles indicator secretions.

  ii. $SO_2$ damages vegetable crops and affects plant growth and nutrient quality of plant, and

  iii. Paper absorbs $SO_2$ which is oxidized to $H_2SO_4$ thus making it brittle and fragile.

**Q 15. What are interference causing agents in sulphur dioxide estimation?**

- Interference causing agents in sulphur dioxide estimation includes $NO_2$, $O_3$, Fe, Mg and V.

**Q 16. For what purpose wet impingers are used?**

- Wet impingers are used for sampling atmospheric $SO_2$, $NO_x$ and other gases for their estimation.

**Q 17. How $NO_2$ and $O_3$ interference in $SO_2$ estimation are removed?**

- Interference of $NO_2$ and $O_3$ in $SO_2$ estimation are removed by adding concentrated sulphuric acid.

**Q 18. How iron, manganese and vanadium interference are removed?**

- Interference of Fe, Mg and V are removed by adding di-sodium salt of EDTA.

**Q 19. What is the principle of sulphur dioxide estimation?**

- Atmospheric sulphur dioxide is absorbed from ambient air into KTCM absorbing media resulted into formation of dichlorosulphitomercurate complex. This complex reacts with para-rosaniline and formaldehyde to form intensely coloured para-rosaniline methyl sulfonic acid. This is measured spectrophotometrically at 560 nm.

**Q 20. What is the molecular weight of sulphur dioxide?**

- The molecular weight of sulphur dioxide is 64.06.

**Q 21. What is the NAAQS for industrial, residential and other area that is ecologically sensitive as notified by Central Government for $SO_2$, for 24 hrs?**

- The NAAQS of $SO_2$ for industrial, residential and other area, which is ecologically sensitive as notified by Central Government is 80 µg/m$^3$ for 24 hours.

**Q 22. What are the standard limits for $SO_2$ concentration for annual average?**

- The standard limits for $SO_2$ concentration for annual average for industrial, residential, rural and other areas is 50 µg/m$^3$ and for ecologically sensitive area it is 20 µg/m$^3$.

**Q 23. What is the sampling frequency for atmospheric sulphur dioxide?**

- The sampling frequency for atmospheric sulphur dioxide for regular monitoring of ambient air quality includes measurement for 24 hrs twice a week, making 104 samples in a year (52 weeks in a year × 2 samples per week = 104 samples).

**Q 24. What are the sampling site selection criteria?**

- The site selection criteria includes:
  - i. The sample collected must be representative in term of time and location.
  - ii. The sample volume should be large enough to perform accurate analysis.
  - iii. The sampling rate must be such that it provides maximum efficiency of collection, and
  - iv. The duration of sampling and frequency of sampling should be such that it records any fluctuation in pollution level.

**Q 25. What are the effects of Sulphur dioxide on Human being?**

- The effects of Sulphur dioxide on Human being includes respiratory tract irritation, diseases like cancer, heart diseases and diabetes mellitus. It causes nasal discharge, sneezing, cough, dyspnoea and post nasal sticky discharge.

**Q 26. What are the effects of Sulphur dioxide on vegetation?**

- Sulphur dioxide damages vegetable crops, affects plant growth and nutrient quality of plant products. Acute exposures to high levels of $SO_2$ kills leaf tissues, causing leaf necrosis. Chronic exposure to plant causes bleaching of leaf pigments. High levels of $SO_2$ affect stomata pore, chloroplast and rate of transpiration.

**Q 27. What are the effects of Sulphur dioxide on materials?**

- Sulphur dioxide attacks on marble, limestone, electrical contact, paper, textile and building. Leather has much affinity toward $SO_2$ which affect its strength and causes it to disintegrate. It accelerates the corrosion rate of metal such as iron, steel, zinc and copper.

**Q 28. What is the contribution of Sulphur dioxide in all pollutants?**

- Sulphur dioxide accounts for about 29% of the total weight of all pollutants.

**Q 29. Why should sulphur dioxide sampled in an absorbing media be analyzed within 48 hours?**

- Sulphur dioxide sampled in an absorbing media should be analyzed within 48 hours because after this it get lost at a rate of 9% per day from absorbing media.

**Q 30. What is the diameter of Midget Impinger orifice?**

- The diameter of Midget Impinger orifice is of 1 mm.

**Q 31. On which factors efficiency of absorbance depends upon.**

- Efficiency of absorbance is dependent upon:
  - i. Amount of surface contact between gas and liquid
  - ii. Contact time
  - iii. Concentration of absorbing media, and
  - iv. Speed of reaction between absorbent and gas.

**Q 32. What is the concentration of standard sodium metabisulfite solution used in $SO_2$ estimation?**

- Concentration of standard sodium metabisulfite solution in $SO_2$ estimation is 300-400 µg of $SO_2$ per mL.

**Q 33. What quantity of absorbing media is taken in an impinger for $SO_2$ sampling?**

- For sampling ambient atmospheric $SO_2$, 30 mL of absorbing media in an impinger should be taken.

**Q 34. What is the absorbance efficiency of $SO_2$ in absorbing media?**

- Absorbance efficiency of $SO_2$ in an absorbing media is 99%.

**Q 35. What is the slope value of standard calibration curve of $SO_2$?**

- The slope value of standard calibration curve of $SO_2$ is about 71.40, this is always greater than $NO_X$.

## ■ 1.4 Oxides of Nitrogen

**Q 1. What does $NO_x$ represents?**

- $NO_X$ represents oxides of nitrogen.

**Q 2. Which instrument is used to sample atmospheric $NO_2$ gases?**

- High volume air sampler is used to sample atmospheric $NO_2$ gases.

**Q 3. What is HVS?**

- HVS is a basic instrument used to sample atmospheric particulate matter and gaseous pollutants.

**Q 4. What is the principle involved in $NO_2$ measurement?**

- Nitrogen dioxide is collected by absorbing ambient air through an absorbing media of sodium hydroxide and sodium arsenite to form a stable solution of sodium nitrite. The nitrite ion produced in a sample is reacted with phosphoric acid, sulphanilamide and N-(1-naphthyl)-ethylenediamine di-hydrochloride (NEDA) to form an azo dye which is measured spectrophotometrically at 540 nm.

**Q 5. Which reagents are used to prepare absorbing media for $NO_2$?**

- Sodium hydroxide, sodium arsenite and N-butyl alcohol are used to prepare absorbing media for $NO_2$.

**Q 6. Which impinger is used to sample atmospheric $NO_2$?**

- Midget impinger is used to sample atmospheric $NO_2$.

**Q 7. Which method is used to determine atmospheric $NO_2$?**

- Modified Jacob and Hochheiser method is used to determine atmospheric $NO_2$ concentration.

**Q 8. What is the time weighted average to determine atmospheric $NO_2$?**

- The time weighted average to determine atmospheric $NO_2$ is 24 hours or 8 hours or hourly monitored values.

**Q 9. Which complex is formed during chemical reaction of $NO_2$ determination?**

- During chemical reaction azo dye complex is formed in $NO_2$ determination.

**Q 10. Which reagent is used to make up the final volume of aliquot?**

- Absorbing media is used to make up the final volume of aliquot.

**Q 11. How are dilutions made for preparation of a calibration curve?**

- Absorbing media is used to make dilutions of a calibration curve.

**Q 12. State the volume of diazotizing reagent added into aliquots.**

- 10 mL of diazotizing reagent is added into aliquots.

**Q 13. At which wavelength $NO_2$ is measured?**

- $NO_2$ is measured at 550 nm on spectrophotometer.

**Q 14. What increases the absorption capacity of an absorbing media?**

- Sodium arsenite and n-butyl alcohol increases the absorption capacity of absorbing media up to 80%, which was previously 30% only.

**Q 15. Which reagents are used for the determination of atmospheric $NO_2$?**

- Reagents used for the determination of $NO_2$ are:
  i. Absorbing media
  ii. Hydrogen peroxide solution
  iii. Sulphanilamide solution (Di-azotizing agent)
  iv. Coupling reagent N-(1-Naphthyl)-ethylenediamine di-hydrochloride, and
  v. Nitrogen dioxide standard solution.

**Q 16. What is a Diazotizing reagent?**

- Diazotizing reagent is a sulphanilamide solution.

**Q 17. What is a Coupling reagent?**

- Coupling reagent is NEDA [N-(1-Naphthyl)-ethylenediamine di-hydrochloride].

**Q 18. Which formula is used to calculate $NO_2$ concentration?**

- For calculation of $NO_2$ concentration the formula used is:

$$NO_2\ (\mu g/m^3) = \frac{\text{Conc. of } NO_2 \text{ in sample} \times 1000}{\text{Total volume of air pass} \times 0.82} \quad \text{OR}$$

$$NO_2\ (\mu g/m^3) = \frac{\text{Conc. of } NO_2 \text{ in sample} \times 1000}{\text{Flow rate}\left(\text{in } \frac{\text{lit}}{\text{min}}\right) \times \text{total time} \times 0.82}$$

**Q 19. Which methods are used to determine atmospheric $NO_2$ concentration?**

- Modified Jacob and Hochheiser (Sodium-Arsenite) method and Chemiluminescence method are used to determine atmospheric $NO_2$ concentration.

**Q 20. What are the adverse effects of higher $NO_2$ concentration on human beings?**

- Higher $NO_2$ concentration causes gum inflammation, internal bleeding, lung cancer, pneumonia and oxygen deficiency in human beings.

**Q 23. What are the adverse effects of higher concentration of $NO_2$ on plants?**

- Higher concentration of $NO_2$ damages plant leaves, affect photosynthesis process and causes chlorosis.

**Q 21. State NAAQS w.r.t. $NO_2$ in ecologically sensitive area.**

- NAAQS w.r.t. $NO_2$ in ecologically sensitive area is 30 $\mu g/m^3$ for annual time weighted average; whereas, for 24 hours time weighted average it is 80 $\mu g/m^3$.

**Q 22. Why should 30 minutes time be given to aliquots before measuring their absorbance?**

- Before measuring absorbance 30 minutes time should be given to aliquots for developing colour.

**Q 23. What is 0.82 in the formula for estimation of $NO_2$ concentration?**

- The 0.82 is an efficiency factor of two impingers combined together, i.e. percentage of $NO_2$ gases absorbed from ambient air (82%). The method of connecting two impingers together is termed as bridging technique. The concentration in first impinger is higher as compared with second impinger.

**Q 24. Which instrument is used to determine atmospheric $NO_2$ concentration?**

- Atmospheric $NO_2$ concentration is determine with the help of spectrophotometer.

**Q 25. Which reagent is used as a reference in determination of $NO_2$?**

- Absorbing media is used as a reference in determination of $NO_2$.

**Q 26. In which unit $NO_2$ concentration is reported?**

- $NO_2$ concentration is reported in $\mu g/m^3$.

**Q 27. Which type of impinger is used for absorption of gases in HVS?**

- Midget type of impinger is used for absorption of gases in HVS.

**Q 28. Give the description of midget type impinger.**

- In midget type impinger, air is drawn at a rate of 0.5 L/min through an orifice of 1 mm in diameter and impinges against a flat surface at 5 mm distance at a jet velocity of about 60 m/s.

**Q 29. What is the sampling duration of $NO_2$ through HVS?**

- Sample duration of $NO_2$ through HVS should be collected continuously for 24 hours. However, if $NO_2$ concentration is more duration may alter to 6, 8 or 12 hours. This can be decided after field survey.

**Q 30. Why rotameter is used in HVS?**

- Rotameter is used in HVS to check flow rate of absorbing gases into impinger.

**Q 31. Where are the solution impingers employed?**

- Solution impingers are employed in situation where high atmosphere particulate loading exists.

**Q 32. What is the use of manometer?**

- Manometer is used to monitor flow rate of incoming ambient air into HVS (in $m^3$/min).

## 1.5 Ammonia

**Q 1. Which method is used for determination of ammonia concentration in ambient air?**

- Nesslerization method is used for determination of ammonia concentration in ambient air.

**Q 2. What is the principle involved in Nesslerization method?**

- When air is passed through an absorbing media, $NH_3$ present in ambient air combines with sulphuric acid forming ammonium sulphate $NH_4(SO_4)$ which is stable for 48 hours and measured spectrophotometrically at 410 nm.

**Q 3. What are the interference causing agents in ammonia determination?**

- Interference causing agents in ammonia determination are $Fe^{2+}$, $Cr^{3-}$, Mn and Ca.

**Q 4. What is the formula for ammonia estimation?**

- The formula for ammonia estimation is:

$$NH_3\left(\mu g/m^3\right) = \frac{\text{Conc. of } NH_3 \text{ in sample } (\text{from graph in } \mu g) \times 1000}{\text{Flow rate } (\text{litre per minute}) \text{ in HVS} \times \text{sampling duration } (\text{min.})}$$

**Q 5. Which reagents are used in ammonia determination?**

- Reagents used in ammonia determination includes:

  i. Absorbing media (Sulphuric acid)
  ii. Nessler's reagent and
  iii. Standard ammonium chloride solution.

**Q 6. What are the adverse effects of ammonia on human being?**

- Adverse effects of ammonia on human being include its reaction with mucous lining of throat, corrosive action on skin and mucus membrane.

**Q 7. What are the adverse effects of ammonia on plants?**

- The high concentration of ammonia causes chlorosis and necrosis in plant leaves.

**Q 8. What are the adverse effects of ammonia on materials?**

- Ammonia damages surface coating of paint, fabrics, erodes building and historical monuments.

**Q 9. At what wavelength ammonia concentration is measured?**

- Ammonia concentration is measured with the help of spectrophotometer at 410 nm.

**Q 10. What is used as a reference solution in ammonia determination?**

- Absorbing media is used as the reference solution in ammonia determination.

**Q 11. What is NAAQS of ammonia for industrial, residential, rural and other area and ecologically sensitive area for annual average?**

- The NAAQS of ammonia for industrial, residential, rural and other area and ecologically sensitive area for annual average is 100 $\mu g/m^3$.

**Q 12. What is NAAQS for ammonia for 24 hours?**

- NAAQS for ammonia for 24 hours is 400 $\mu g/m^3$.

**Q 13. In which unit ammonia concentration is reported?**

- Ammonia concentration is reported in $\mu g/m^3$.

**Q 14. Name other methods used for measurement of ammonia concentration.**

- Chemiluminescence and Indophenol blue method are also used for measurement of ammonia concentration.

**Q 15. What is NAAQS for ammonia in ecologically sensitive area for annual average?**

- The NAAQS for ammonia in ecologically sensitive area for annual average is 100 $\mu g/m^3$.

**Q 16. What is NAAQS for ammonia in ecologically sensitive area for 24 hours?**

- The NAAQS for ammonia in ecologically sensitive area for 24 hours is 400 $\mu g/m^3$.

**Q 17. Which chemicals are used for preparation of Nessler's reagent?**

- Mercuric iodide and potassium iodide are used for preparation of Nessler's reagent.

**Q 18. What is the flow rate of atmospheric ammonia into absorbing media?**

- The flow rate of atmospheric ammonia into absorbing media should be 0.5 litres per minute (LPM).

**Q 19. Name the instrument used for sampling atmospheric ammonia.**

- High volume air sampler is used for sampling atmospheric ammonia.

**Q 20. By which instrument ammonia concentration is measured in laboratory?**

- Ammonia concentration in laboratory is measured with the help of spectrophotometer.

**Q 21. Name the sampling apparatus used for absorbing atmospheric ammonia.**

- Impinger is used for absorbing atmospheric ammonia.

**Q 22. How ammonia gets entrapped into absorbing medium?**

- Ammonia gets entrapped into absorbing media by absorption process.

**Q 23. State the significance of 0.5 LPM flow rate in ammonia sampling.**

- The significance of 0.5 LPM flow rate in ammonia sampling is that at this flow rate ammonia gets dissolved completely into absorbing media.

**Q 24. How is the flow rate of atmospheric gases checked/adjusted for ammonia sampling?**

- Flow rate of atmospheric gases for ammonia sampling is checked/adjusted with the help of rotameter.

## 1.6 Sulphation Rate

**Q 1. What is Sulphation rate?**

- Sulphation rate is a cumulative index useful for evaluating sulphur dioxide and other sulphur compounds concentration in atmosphere.

**Q 2. Which method is used for estimation of sulphation rate?**

- Sulphation rate is estimated by lead peroxide candle method.

**Q 3. Why lead peroxide is used in estimation of sulphation rate?**

- Lead peroxide, due to its oxidizing power converts other compound such as mercaptans and hydrogen sulphides into sulphate. It also fixes sulphur trioxide and sulphuric acid mist present in the atmosphere. Thus, lead peroxide is used in estimation of sulphation rate.

**Q 4. Which paste is applied on candles for estimation of sulphation rate?**

- A mixture of 8 g of lead peroxide and 6 mL of gum mucilage (5 g gum tragacanth in 10 mL alcohol and then added 190 mL of distilled water with stirring) is applied on candle for estimation of sulphation rate.

**Q 5. Why barium chloride is used in estimation of sulphation rate?**

- Sulphate ion is precipitated in hydrochloric acid medium with barium chloride so as to form barium sulphate. The absorbance of which is proportional to concentration of the sulphate present. Hence, barium chloride is used in estimation of sulphation rate.

**Q 6. Why is conditioning reagent used in sulphation rate estimation?**

- Conditioning reagent is used in sulphation rate estimation so as to maintain acidic condition to form barium sulphate.

**Q 7. What is the aliquot factor/conversion factor?**

- Aliquot factor or conversion factor for estimation of sulphation rate is 2.425.

**Q 8. Why is the aliquot factor 2.425?**

- Aliquot factor is 2.425 because 1 mg $SO_4$ = 2.425 mg $BaSO_4$.

**Q 9. Why aliquot factor/conversion factor is used in calculation of sulphation rate?**

- Conversion factor is used in calculation of sulphation rate to convert mg of sulphate to mg of $BaSO_4$, thus aliquot factor/conversion factor is used in calculation of sulphation rate.

**Q 10. At what wavelength sulphate concentration is estimated?**

- Sulphate concentration is estimated at 420 nm on spectrophotometer.

**Q 11. State the formula used in estimation of sulphation rate.**

- $$\text{Sulphation rate}\left(\text{mgSO}_4/100\ \text{cm}^2/\text{day}\right)=\frac{\text{Total BaSO}_4\text{due to sulphation}\times 0.348}{\text{No. of exposure days}}$$

**Q 12. What is 0.343 factor in sulphation rate calculation?**

- Factor 0.343 in sulphation rate calculation expresses the relationship, 1 mg $BaSO_4$ = 0.343 mg sulphate.

**Q 13. Which filter paper is used to filter sulphation rate sample?**

- Whatman filter paper 41 is used to filter sulphation rate sample.

**Q 14. Who developed the lead peroxide candle method?**

- Wisden and McConnell (in 1934) developed the lead peroxide candle method.

**Q 15. Why louvered box is used for exposure of candle?**

- Louvered box keeps air off but allows air to come into contact with fabric surface and thus is used for exposure of candle.

**Q 16. Why is conc. HCl used in sulphation rate estimation?**

- Concentrated HCl is used in sulphation rate estimation to neutralize the filtrate to a pH range of 3 to 4 and further to prevent frothing.

**Q 17. What is the limitation of sulphation rate method?**

- The only limitation of sulphation rate method is dust and particulate matter which gets deposited on lead to give erroneous results.

**Q 18. On which factors sulphation rate collection efficiency is depends upon?**

- Sulphation rate collection efficiency depends upon factors such as temperature, humidity and wind speed.

**Q 19. Which factors interferes with qualitative analysis of sulphation rate?**

- Hydrogen sulphide and $H_2SO_4$ interferes in qualitative analysis of sulphation rate.

**Q 20. What is the concentration of standard sulphate solution?**

- Concentration of standard sulphate solution is 1 mL = 0.10 mg $SO_4^{2-}$ = 100 µg $SO_4^{2-}$.

**Q 21. Why filtration is carried out in sulphation rate estimation?**

- Filtration is carried out in sulphation rate estimation is to remove interference of colour and turbidity.

# 2

# WATER

## 2.1 Colour

**Q 1. What is colour in water?**

- The colour in water means true colour, i.e., the colour of water from which turbidity has been removed. True colour is dissolved in nature and can't be removed by filtration or centrifugation.

**Q 2. How colour occurs in water?**

- In ambient stream, water colour may result from the presence of natural metallic ions (iron and manganese), humus and peat materials, planktons, weeds and some industrial wastes.

**Q 3. Why should colour be removed from water?**

- Colour should be removed from water to make water a source suitable for general and industrial applications. Highly coloured industrial effluents may require colour removal before being discharged into normal water courses.

**Q 4. Which components exhibits colour to water?**

- The colouring material results from contact of water with organic debris, such as leaves, needles of conifers wood (in various stages of decomposition), tannins, humic acid, fulvic acid and humates from the decomposition cellulose and lignin.

**Q 5. In which form natural colour exists in water?**

- Natural colour exists in water primarily as negatively charged colloidal particles.

**Q 6. How decolourization of water takes place? And why pH and temperature of water are measured with colour?**

- Usually colour can be readily removed by coagulation with the aid of a salt having trivalent metallic ions such as salts of aluminium and iron. On addition of salts for colour removal, colour intensity increases generally with increase in pH and temperature and thus recording of these parameters are recommended.

**Q 7. What are the types of colours found in water?**

- There are two types of colours found in water. They are true colour and apparent colour.

**Q 8. What is "True Colour" ?**

- True colour is the colour of water from which turbidity has been removed. True colour is non-filterable or rather is dissolved in nature and cannot be removed by filtration or centrifugation. True colour can be caused by vegetable or organic extracts that are colloidal in nature.

**Q 9. What is "Apparent Colour"?**

- The term 'Apparent colour' includes colour due to dissolved and suspended components found in water. In some highly coloured industrial wastewaters colour is contributed principally by colloidal or suspended material. In such cases both true colour and apparent colour should be determined.

**Q 10. How "trihalomethanes" are formed?**

- Water containing colouring matter derived from natural substances undergoing decay in swamps and forests is not considered to possess harmful or toxic properties. This gives a yellowish - brown appearance to water like urine. Disinfection by chlorination of such water containing natural organics may result in the formation of a class of organohalides known as trihalomethanes.

**Q 11. What is the minimum standard of colour in public water supplies? Why?**

- The minimum standard of colour in public water supplies intended for human use should not have a true colour exceeding 10 chloroplatinate units (CPUs). Public health officials are aware of the fact that consumers will seek other sources of drinking water if the public water supply is not aesthetically acceptable, no matter how safe the water may be from a hygienic viewpoint.

**Q 12. Which method is used for measurement of colour?**

- Visual comparison method is applicable to nearly all samples of potable water for measurement of colour.

**Q 13. What is Platinum-Cobalt method?**

- Platinum-Cobalt method is useful for measuring colour in water derived from naturally occurring materials.

**Q 14. What is CPU (Chloroplatinate unit)?**

- One unit of colour is defined as that colour produced by 1.0 mg/L of platinum complex in the form of chloroplatinate ion and is called as CPU (Chloroplatinate unit).

**Q 15. What are the interferences in colour measurement?**

- The interferences in colour measurement are
  a) Turbidity in excess of 5 NTU may seriously affect colour of sample.
  b) Colour is extremely pH dependent and invariably increases as pH of sample increases.

**Q 16. Which reagents are used for the preparation of standard colour solution? How it is made?**

- Compounds used for preparation of standard colour solution are: a) $K_2PtCl_6$ (for platinum metallic) and b) $CoCl_2.6H_2O$ (for metallic cobalt). Dissolve 1.246 g $K_2PtCl_6$ (equivalent to 500 mg metallic platinum) and 1 g crystallized $CoCl_2.6H_2O$ (equivalent to about 250 mg metallic cobalt) in distilled water that has been filtered to remove any slight turbidity. Then add 100 mL conc. HCl and dilute with distilled water to 1000 mL. This stock standard colour solution has a concentration of 500 CP units.

**Q 17. What would be the final colour of 25 mL sample that was diluted to the final volume of 50 mL when the colour comparison test showed 70 CPU's on this diluted sample?**

- Colour unit = $A \times 50 / B$

Where, A = estimated colour of a diluted sample =?

B = mL of sample taken for dilution = 25 mL

Colour unit = 70 CPU

$$A = \frac{\text{Colour unit} \times B}{50}$$

$$A = \frac{70 \times 25}{50} = 35 \text{ Hazen unit}$$

**Q 18. What is the principle of platinum cobalt method for estimation of colour?**

- Colour formed due to combination of platinum in the form of chloroplatinic acid and cobaltous chloride is visually compared and is expressed in terms of Hazen unit.

**Q 19. What is the wavelength range in photoelectric method for colour measurement?**

- The wavelength range used for colour measurement is between 385 nm and 470 nm in photoelectric method.

**Q 20. What will be the reference for spectrophotometer for photoelectric method in colour determination?**

- The reference for spectrophotometer for photoelectric method in colour determination is a 10 mm cell with distilled water in reference cell.

## ■ 2.2 Turbidity

**Q 1. Define turbidity.**

- Turbidity is defined as the optical property of a suspension which causes light to be scattered rather than transmitted through a sample. The interactions in any natural suspension are complex and the term turbidity is frequently used in a qualitative sense.

**Q 2. What is Rayleigh theory for scattering?**

- For particles with diameters smaller than about 1/20 of the wavelength (L) of incident light, the Rayleigh theory of scattering indicates that the intensity of the scattered light (Is), is related to the particle volume (V), the number (N) of particles, and the wavelength (L) by the relationship $Is = (V^2/L^4)\ (N)$.

**Q 3. How intensity of scattered light (Is) varies with particle size?**

- When particle size increases above a diameter of I/20 (L), the intensity of scattered light becomes a complex function of particle size and angle of emission. For constant number of particles and a given wavelength increase to a maximum and decreases as particle size increases.

**Q 4. Which are the different components contributing to the turbidity of water?**

- The turbidity of water sample is the reduction of transparency due to the presence of suspended particulate matter. Such material may consists of clay or silt, finely divided organic matter, plankton, or other microscopic organisms which have caused light to be scattered and absorbed rather than transmitted.

**Q 5. Which instrument is used to measure turbidity?**

- Nephelometer/Turbidimeter is used to measure the turbidity of water sample.

**Q 6. What are the causes of turbidity in water?**

- Turbidity is caused by wide variety of suspended matter in water which ranges in size from colloidal to coarse dispersions, depending upon the degree of turbulence and also ranges from pure inorganic substances to those that are highly organic in nature.

**Q 7. What are the problems associated with turbid water?**

- Turbid water is undesirable from aesthetic point of view in drinking water supplies and may also affect products in industries. Turbid water poses a number of problems in water treatment plants.

**Q 8. Which methods are used for measurement of turbidity?**

- There are two methods to measure turbidity of water sample: Turbidimetry and Nephelometry.

**Q 9. What is Turbidimetry?**

- When turbidity is measured by its effect on transmission of light then it is termed as turbidimetry. Turbidimetry can be used for samples with moderate turbidity.

**Q 10. What is Nephelometry?**

- When turbidity is measured by its effect on scattering of light then it is termed as nephelometry. Nephelometry can be used for sample with low turbidity. Higher the intensity of scattered light higher is the turbidity.

**Q 11. What is the importance of aesthetics in public water supplies?**

- Consumers of public water supplies expect and have a right to demand turbidity free water. Lay people are aware that domestic wastewater is highly turbid. Any turbidity in drinking water can be associated with possible wastewater pollution and health hazards by it.

**Q 12. What is the role of filterability and disinfection in removal of turbidity?**

- *Filterability:* Satisfactory operation and rapid sand filter depends upon effective removal of turbidity by chemical coagulation before water is admitted to the filter. Failure to do so results into short filter run and production of inferior quality water unless, filters of special construction are used.

  *Disinfection:* In turbid waters, most of the harmful organisms are exposed to the action of chemical disinfection. However, cases in which turbidity is caused by organic materials instead of inorganic clay or silt, many are encased in particles, in particularly if the organic material solids are of a spent water origin, and protected from the disinfectant chemicals.

**Q 13. What was the original standard unit for turbidity?**

- The original standard unit chosen for turbidity was 1 mg $SiO_2$/L= 1 unit of turbidity.

**Q 14. What was the previous method for turbidity determination?**

- The previous method for determination of turbidity was Jackson candle turbidimeter.

**Q 15. What was used to calibrate Jackson candle turbidimeter?**

- Standard suspension of pure silica was used to calibrate Jackson candle turbidimeter.

**Q 16. What is used as a standard solution in turbidimetry?**

- Formazin solution is used as a standard in turbidimetry.

**Q 17. State the applications of turbidity data.**

- Turbidity measurement is of particular importance in the field of water supply. It has limited use in the field of domestic and industrial waste treatment.

  Water supplies obtained from river usually require chemical flocculation because of high turbidity. Turbidity measurements are used to determine the effectiveness of the treatment produced with different chemicals and the dosages needed.

  Turbidity measurement helps to gauge the amount of chemicals needed from day-to-day in the operation of treatment works. Turbidity measurement can be used to advantage because of the speed with which it can be made, to gain the necessary information. By their use, chemical dosages can be adjusted to use the minimum amount of chemicals while producing a high-quality effluent.

**Q 18. In which unit turbidity is reported?**

- Turbidity is reported in NTU (Nephelometric Turbidity Unit).

## 2.3 pH

**Q 1. What is pH?**

- pH is the power of hydrogen ion concentration and it is used to measure acidity and/or basicity of a solution.

**Q 2. Who discovered the pH scale?**

- The pH scale was discovered by Danish chemist, P.L. Sorenson in 1909 at Carlsberg laboratory.

**Q 3. Which types of electrodes are used in measuring pH of a solution?**

- An indicator electrode (an electrode responsive to hydrogen ions such as glass electrode) and a reference electrode (usually a mercury calomel electrode) are used to measure pH of a solution.

**Q 4. What are the components of glass electrode?**

- The glass electrode consists of a glass shaft on which a bulb of special glass is mounted.

**Q 5. Which internal solution is filled in glass electrode?**

- Potassium chloride (KCl) of 0.1 mol/L is filled in glass electrode as a internal solution.

**Q 6. Who provides electrical contact in glass electrode?**

- Electrical contact in glass electrode is provided by silver wire immersed in KCl solution.

**Q 7. What is the normal range of pH value on a pH meter?**

- The normal range of pH value on a pH meter is 0 - 14.

**Q 8. What is the acidic range of pH scale?**

- The acidic range of pH scale is 0 - 6.9.

**Q 9. What is the basic range of pH scale?**

- The basic range of pH scale is 7.1 - 14.

**Q 10. What is the neutral point of the pH scale?**

- The neutral point on the pH scale is 7.0.

**Q 11. What is a buffer solution?**

- A buffer solution is a solution that resists changing in its pH caused by addition of small quantity of acid or alkali.

**Q 12. How does a buffer solution work to keep the pH constant?**

- Buffer solution act as a neutralizing agent that contains both positive and negative ions of weak acid and weak base.

***Example : 1.*** If a base such as NaOH is added to the solution, the positive $H^+$ ions will react with the negative $(OH^-)$ to neutralize the solution.

***Example : 2.*** Similarly, if an acid is added to the buffer, the negative ions will react with hydronium ions to neutralize the solution.

**Q 13. What is meant by acidic buffer?**

- A weak acid and its conjugate base is known as acidic buffer.

**Q 14. What is meant by basic buffer?**

- A weak base and its weak conjugate acid is known as basic buffer.

**Q 15. Why buffer solutions are used to calibrate pH meter and not diluted solution?**

- Buffer solutions are used to calibrate pH meter because un-buffered solution will change pH with exposure to atmospheric $CO_2$.

**Q 16. Why pH scale ranges from 0-14 only at 25°C ?**

- The pH scale is based on the dissociation constant of water, which has a value of $1 \times 10^{-14}$ at 25°C. The negative logarithm to the base 10 of this value of $-\log_{10}(1 \times 10^{-14}$ at 25°C) is 14, because of this the pH scale ranges from 0-14 only at 25°C.

**Q 17. What is the role of litmus paper in measuring pH?**

- The role of litmus paper in measuring pH is to provide a numerical value of the pH.

**Q 18. What is the relation between pH and temperature?**

- The relationship between pH and temperature is that, if in a weak acid or weak base, most of the acid or base is not ionized, the energy at high temperature helps these molecules to break them. In this case an acid produces more $H^+$ ions and in case of bases it produces more $OH^-$ ions.

**Q 19. Why is reference electrode used in measuring pH?**

- Reference electrode is used in measuring pH because it provides an accurate pH measurement, for that reference must have constant and stable potential.

**Q 20. Is pH the measurement of hydrogen ion concentration or ion activity?**

- pH is the measure of hydrogen ion activity.

**Q 21. What does pH scale measures?**

- The pH scale measures the acidity or alkalinity of a solution.

**Q 22. What is the pH of rain water?**

- The pH of rain water is slightly acidic (5.4), because as rain drop falls on the earth surface from atmosphere, the atmospheric $CO_2$ gets dissolved into rain droplets and forms a weak carbonic acid.

**Q 23. How can one increase or decrease the pH of water?**

- By adding a few drops of an alkali to water, its pH can be increased and by adding a few drops of an acid to water the pH of water can be decreased.

**Q 24. What are indicators?**

- Indicators are chemicals that show whether a given solution is acidic or basic by sudden change of colour.

**Q 25. What is an Universal indicator? What are its advantages?**

- A universal indicator is a mixed indicator of organic chemicals which not only shows whether the given solution is acidic or basic but also shows the approximate pH values by giving a wide particular colour for a specific value of pH.

**Q 26. What is the standard value for pH in drinking water?**

- According to IS 10500:2012, 6.5-8.5 is the standard value for pH in drinking water.

**Q 27. In which water, the hydrogen and hydroxide ions concentration remains stable?**

- In pure water, the hydrogen and hydroxide ions concentration are present in a 1:1 ratio so, they remain stable.

**Q 28. What are acids and bases?**

- a) Acids are the substances in which the concentration of hydrogen ions increases when they are placed in water.

  b) Bases are the substance in which the concentration of hydroxide ions increases when they are placed in water.

**Q 29. What are strong acids and strong bases?**

- Strong acids are those which are completely ionized when dissolved in water and strong bases are those which are completely ionized when dissolved in water.

**Q 30. What are weak acids and weak bases?**

- Weak acids are those which are not completely ionized when dissolved in water and weak bases are those which are not completely ionized when dissolved in water.

**Q 31. Which are the three common indicators used in laboratory?**

- Litmus test, methyl orange and phenolphthalein are the three common indicators used in laboratory.

**Q 32. Name the methods used to measure pH of a solution.**

- pH meter and universal indicator paper are the methods used to measure pH of a solution.

**Q 33. Which is the standard method used to measure pH of a solution?**

- Electrometric method is the standard method used to measure pH of a solution.

**Q 34. What is the physiological value of pH?**

- The physiological value of pH is 7.4.

**Q 35. What is the hydrogen ion concentration of a solution at physiological pH?**

- -7.4 M is the hydrogen ion concentration of a solution at physiological pH.

**Q 36. What is the hydroxyl ion concentration of a solution at physiological pH?**

- 8.6 M is the hydroxyl ion concentration of a solution at physiological pH.

**Q 37. What is the relationship between $H^+$ ions and $OH^-$ ions concentration?**

- The relationship between $H^+$ and $OH^-$ ions is $Kw=[H^+]\,[OH^-]$, $Kw=1\times10^{-14}$.

**Q 38. In pH why p is written in small and H is written in capital letter.**

- In pH, p stands for negative logarithm and H stands for hydrogen ions concentration that's why in pH, p is written in small and H is written in capital letter.

**Q 39. What is zero on pH scale?**

- Zero represents a very strong acid on pH scale.

**Q 40. On which law pH experiment is based upon?**

- The pH experiment is based upon the law of mass action.

**Q 41. Which oxides show both acidic and basic properties?**

- An amphoteric oxide shows both acidic and basic properties.

**Q 42. What are the components of pH indicator paper?**

- A pH indicator paper contains halochromic chemical compounds that are added in small amount to the solution so that the pH (acidity or basicity) of the solution can be determined visually.

**Q 43. What is meant by universal indicator?**

- The indicator which changes over a pH value range of 0-14 to indicate acidity or basicity of a solution is known as universal indicator.

**Q 44. Give cell schematic representation of pH meter.**

- Cell schematic representation of pH meter is as follows:

  Glass electrode || Reference Solution || **Test Solution** || Glass electrode

  Ag(s) | AgCl(s) | KCl(aq) || $1\times10^{-7}$ M $H^+$ solution || glass membrane || **Test Solution** || junction || KCl(aq) | AgCl(s) | Ag(s)

**Q 45. How should pH electrode after use be stored?**

- pH electrode after use should be stored in a solution of its own ion. It is necessary to prevent the glass membrane from drying out because the performance is dependent on the existence of a hydrated layer, which forms slowly.

**Q 46. What is the modification in a typical modern pH electrode?**

- A typical modern pH electrode is a combination electrode, which combines both the glass and reference electrodes into one body, which was previously, used to be two separate electrodes.

## 2.4 Conductivity

**Q 1. What is Conductivity?**

- Conductivity is the capacity to carry an electric current.

**Q 2. What is Conductivity meter?**

- The meter or instrument designed to measure an electrical conductivity is known as conductivity meter.

**Q 3. Who discovered conductivity meter?**

- Alessandro Volta an Italian Physicist discovered conductivity meter.

**Q 4. What is an Electrical conductivity?**

- Electrical conductivity refers to the movement of charged particles through matter.

**Q 5. State methods to measure an electrical conductivity.**

- There are two methods to measure an electrical conductivity they are 1) Steady-state method and 2) Non-steady state method.

**Q 6. What is the standard method for measuring conductivity?**

- Conductivity meter method is the standard method for measuring conductivity.

**Q 7. What is the unit of conductivity?**

- The unit of conductivity is µSiemens/cm or µmhos/cm.

**Q 8. Which solutions have lower electrical conductivity?**

- Molecular solutions have lower electrical conductivity because they do not contain any metallic ions.

**Q 9. Which solutions have higher electrical conductivity?**

- Metallic solutions have higher electrical conductivity because they contain maximum number of metallic ions.

**Q 10. What is electrical resistivity?**

- Electrical resistivity is the property to oppose the free flow of electrons in electric current.

**Q 11. What is meant by conductive polymer?**

- Conductive polymers are those that conduct electricity.

**Q 12. Which type of chemicals are used for making of conductive polymers?**

- Poly acetylene and poly vinylene are used for making conductive polymers.

**Q 13. Which is the most conductive metal?**

- Silver is the most conductive metal.

**Q 14. Which metal is least conductive?**

- Copper is the least conductive metal.

**Q 15. What is mean by Thermal conductivity?**

- Thermal conductivity is the property of a material or ability to conduct heat is known as Thermal conductivity.

**Q 16. What is the conductivity of distilled water?**

- Distilled water should not have any conductivity or have minimum conductivity of 0.5-3 μmhos/cm.

**Q 17. By which material conductivity electrode is made up of?**

- The conductivity electrode is made up of platinum metal.

**Q 18. On which law conductivity experiment is based upon?**

- The conductivity experiment is based on Kohlrausch's law.

**Q 19. What is the difference between specific conductance and conductivity?**

- The difference between specific conductance and conductivity is that, the specific conductance is the scientific term for conductivity values that have been compensated to 25°C.

**Q 20. How rough estimation of dissolved solids can be carried out with the help of specific conductance?**

- Rough estimation of dissolved solids present in water sample can be made by multiplying specific conductance by an empirical factor which may vary from 0.55 to 0.90 depending on the soluble components of water and on the temperature of measurement.

**Q 21. What are the components of conductivity meter?**

- Most of the instruments commercially available for measurement of conductivity consists of a source of an alternating current, a Wheatstone bridge–a null indicator and a conductivity cell consisting of a pair of rigidly mounted electrodes.

**Q 22. What is a cell constant of a conductivity cell?**

- Cell constant of a conductivity cell is the product of measured resistance of the cell and the conductivity. Cell constant of each cell is different and mentioned on the cell.

**Q 23. How cell constant can be determined?**

- Cell constant can be determined by using standard solution of KCl (0.01 M).

**Q 24. What is the alternative method for estimation of conductivity?**

- Conductivity can alternatively be estimated by comparison with a cell of known cell constant.

**Q 25. What is the conductivity of 0.01 M KCl solution?**

- The conductivity of 0.01 M KCl solution is 1411.8 μmhos/cm.

**Q 26. What does 'S' stands for in conductivity unit?**

- 'S' in conductivity unit stands for Siemens.

**Q 27. What is Siemens?**

- Siemens is a SI unit of electrical conductance equal to the conductance of a circuit or element that has a resistance of 1 ohm. $1S = 10^{-1}\ \Omega$.

**Q 28. What Siemens was formerly called as?**

- Siemens was formerly called as mho or reciprocal of ohm.

**Q 29. After whom the Siemens is named as.**

- Siemens is named after the German born British engineer Sir William Siemens (1823-83).

**Q 30. What is the relation between 1 μS/cm and mS/cm?**

- 1μS/cm = 0.001 mS/cm = 0.000001 S/cm = 1μmho/cm.

**Q 31. Why formerly conductivity was reported in μmhos/cm and now as μS/cm?**

- The conductivity was formerly measured in μmhos/cm. As a result of change in metric system, the unit now is reported in μS/cm. 1μmho/cm = 1μS/cm.

**Q 32. How to get the value of μS/cm into ppm?**

- To get a value of μS/cm into ppm, multiply it with 0.5.

**Q 33. How to get the values of mS/cm into μS/cm?**

- To get a value of mS/cm to μS/cm, multiply it with 1000.

**Q 34. Explain cell constant.**

- Suppose two electrodes are placed at a distance of l cm and having a cross section area of a sq. cm. then cell constant ($x$) is given by $x$ = l/a. Hence, the ratio of l/a is known as cell constant.

**Q 35. What is specific conductance?**

- The specific conductance of a conductor is the reciprocal of specific resistance and is generally denoted by K (Greek, small Kappa).

$$R = 1/K\ l/a\ \text{ohm}$$

$$K = 1/R * l/a\ \text{ohm}^{-1}\ \text{cm}^{-1}$$

$$\text{If } l = 1\ \text{cm, and a} = 1\ \text{sq. cm.}$$

$$K = 1/R$$

$$K = 1/\rho$$

$$\rho = R$$

$$1/R = 1/\rho$$

$$\text{Since } K = 1/\rho = \frac{1}{\mathit{omh\ cm}}$$

$$= \mathit{ohm}^{-1}\ \text{cm}^{-1} = \mathit{mho}/\text{cm}$$

**Q 36. What is specific resistance?**

- The resistance of a uniform column of the material of the conductor having a length of 1 cm and a cross section of 1 sq cm. $\rho = R\ a/l$ = ohm (cm)$^2$/cm = ohm. cm

Where, $\rho$ = Specific resistance
R = Resistance of conductor
a = Area of cross section of a conductor, and
$l$ = Length of a conductor.

**Q 37. Variation in conductivity is observed with respect to.**

- Variation in conductivity is observed with respect to number and types of ions in solution.

**Q 38. For continuous monitoring and field studies which material should the electrode be made up of?**

- For continuous monitoring and field studies, electrode made of durable common metal such as stainless steel should be used.

**Q 39. State various conversion formulas for converting conductivity.**

- Conductivity conversion:

| Multiply to get → | ← To get divide |
|---|---|
| µS/cm | 1 µmho/cm |
| mS/cm | 1000 µS/cm |
| µS/cm | 0.5 ppm |

**Q 40. How well a solution conducts electricity depends on which factors?**

- A solution's electrical conductivity depends on factors such as concentration, mobility of ions, valence of ions and temperature.

## ■ 2.5 Solids

**Q 1. What are solids?**

- Solids are the matter either filterable or non-filterable that remains as a residue upon evaporation and subsequent drying at a defined temperature (103°C-105°C).

**Q 2. What are the sources of solids?**

- The sources of solids in water includes natural sources, sewage, urban run-off, industrial wastewater, chemicals used in water treatment processes and plumbing material.

**Q 3. State types of solids.**

- The types of solids includes total solids, suspended solids, dissolved solids, volatile solids, organic solids and inorganic solids.

**Q 4. What are the adverse effects of high solids on potable water?**

- Water having high solids concentration often has an odd taste and odour in a potable water source.

**Q 5. What do total solids represent?**

- Total solids represent both suspended solids and dissolved solids. It is also the solids left after a known volume of sample is evaporated at 103°C-105°C.

**Q 6. What are total suspended solids?**

- Total suspended solids are the portion of the total solids retained by a filter paper. It is also called as filterable solids.

**Q 7. On which principle estimation of total suspended solids is based upon?**

- Estimation of total suspended solids is based upon gravimetric analysis.

**Q 8. What are dissolved solids?**

- Dissolved solids are solids left, that come through from the suspended solids analysis after evaporation of a known volume of filtrate. They are also called as non-filterable solids.

**Q 9. At what temperature, suspended solids collected on filter paper is dried in an oven?**

- Suspended solids collected on filter paper is oven dried at a temperature of 103°C -105°C.

**Q 10. What volatile solids determination indicates?**

- Volatile solids determination indicates amount of organic matter present in suspended solids.

**Q 11. At what temperature estimation of volatile solids is carried out?**

- Estimation of volatile solids is carried out at a temperature of 550°C in a muffle furnace. At this temperature organic matter, particularly carbon residue can be oxidized at reasonable speed. At this temperature decomposition of inorganic salts is minimum. Ammonium compounds are volatilized and most other inorganic salts are relatively stable.

**Q 12. State the formula for calculation of total suspended solids in mg/L.**

- Total suspended solids (mg/L) = $(W_2 - W_1) \times (1000 \text{ mg/g}) \times (1000 \text{ mL/L})/$ volume of sample used (mL). Where, $W_2$ is final weight of filter paper and $W_1$ is initial weight of filter paper.

**Q 13. Which apparatus is used to measure settleable solids?**

- Imhoff cone is used to measure settleable solids.

**Q 14. What is the capacity of imhoff cone?**

- The capacity of imhoff cone is 1000 mL.

**Q 15. Which filter paper is used in estimation of suspended solids?**

- Glass fibre circular filter paper having 47 mm diameter size and 0.45 micron pore size is used for estimation of suspended solids.

**Q 16. Give the composition of dissolved form in potable water.**

- Dissolved form in potable water mainly consists of inorganic salts, small amount of organic matter and dissolved gases.

**Q 17. What is the range of dissolved solids concentration in potable water?**

- The total dissolved solids concentration of potable water usually ranges from 20 to 1000 mg/L. As hardness increases total dissolved solids concentration also increases.

**Q 18. At what temperature total dissolved solids estimation is carried out.**

- Total dissolved solids estimation is carried out at 180°C. This elevated temperature removes all mechanically occulted water.

**Q 19. What are undissolved substances referred as?**

- Undissolved substances are usually referred as suspended matter or suspended solids.

**Q 20. At what temperature, calcium carbonate present in volatile solids analysis is decomposed?**

- Calcium carbonate present in volatile solids is decomposed at 825°C.

**Q 21. Which solids will settle under gravitational force as settleable solids?**

- Coarse suspended solids having specific gravity sufficiently greater than that of water will settle due to influence of gravity in quiescent condition.

**Q 22. What is sludge?**

- Sludge is accumulation of settleable solids.

**Q 23. State application of determination of suspended and volatile suspended solids in water samples.**

- Suspended and volatile suspended solids determination are used to evaluate the strength of domestic and industrial waste.

## 2.6 Acidity

**Q 1. Define acidity.**

- Acidity is defined as the capacity to neutralize alkaline solutions.

**Q 2. How acidity is expressed?**

- Acidity is expressed in terms of mg/L as $CaCO_3$.

**Q 3. What are interference causing agents in acidity analysis?**

- Residual chlorine, colour, turbidity, iron, aluminium etc. are common interference causing agents in acidity analysis.

**Q 4. How residual chlorine can be removed in acidity?**

- Residual chlorine can be removed by addition of calculated amount of sodium thiosulphate solution while other can be avoided by carrying out potentiometric titration.

**Q 5. How acidity can be calculated in terms of mg/L as $CaCO_3$?**

- Acidity as $CaCO_3$ mg/L can be calculated by mg/L = mL (NaOH MO) + mL (NaOH P) × N × 50 × 1000/mL of sample. Where, MO = methyl orange and P = Phenolphthalein indicator.

**Q 6. What is the main source of acidity?**

- Dissolved carbon dioxide is the main source of acidity.

**Q 7. Which titrant is used for acidity analysis?**

- Standard sodium hydroxide (0.02 *N*) is used as a titrant for acidity analysis.

**Q 8. Which indicators are used to determine acidity?**

- Indicators used for determination of acidity are Phenolphthalein and Methyl orange.

**Q 9. What is the colour change at the end point of acidity?**

- Red purple colour shows the end point of acidity.

**Q 10. In acidity, which solution is used for standardization?**

- In acidity, standardization is carried out by Sodium hydroxide solution.

**Q 11. For standardization of Sodium hydroxide which titrant is used in acidity?**

- For standardization of sodium hydroxide the titrant used in acidity is potassium hydrogen phthalate.

**Q 12. Which indicator is used for standardization of Sodium hydroxide solution?**

- Methyl orange indicator is used for standardization of Sodium hydroxide solution.

**Q 13. How $CO_2$ free distilled water is prepared?**

- Carbon dioxide free distilled water is prepared by boiling and cooling distilled water. While cooling it, it has to be covered with proper cover so as to avoid entry of atmospheric $CO_2$ into it.

**Q 14. How normality of NaOH is calculated in standardization procedure of acidity?**

- Normality of NaOH in standardization procedure of acidity is calculated by standardizing it against 0.02 *N* potassium hydrogen phthalate.

**Q 15. What is acidity?**

- Acidity of a solution is its capacity to donate $H^+$ ions.

**Q 16. How is water sample buffered?**

- Most natural water and sewage water are buffered by $CO_2$-bi-carbonate system. Acidity present due to free $CO_2$ has no significance from public health view point.

**Q 17. How does water contains mineral acidity?**

- Water contains mineral acidity due to presence of sulphuric acid.

**Q 18. Which acids are acceptable in acidity?**

- Hydrochloric acid and Nitric acid are acceptable in acidity.

**Q 19. How mineral acids can be calculated?**

- Mineral acids present in a solution can be calculated by titrating or neutralizing sample to pH 4.

**Q 20. What problems are caused by acidic water?**

- Acidic water causes corrosion of metals and interferes in water softening.

**Q 21. In acidity which elements are neutralized completely?**

- Carbon dioxide and bi-carbonates (carbonic acid) present in sample can be neutralized completely by continuing titration to pH 8.3 with the help of 0.02 *N* NaOH.

**Q 22. How acidity can be determined?**

- Acidity can be determined by measuring a volume of standard base (0.02 *N* NaOH) to be added to raise the pH to a specific value.

**Q 23. Which constituents are present in acidity?**

- Acidity is a net effect of presence of several constituents, including dissolved carbon-dioxide, dissolved multivalent metal ions, strong mineral acids such as sulphuric, nitric and hydrochloric acid and weak organic acids such as acetic acid.

**Q 24. Which water contains acidity?**

- Water having a pH less than 8.5 contains acidity.

**Q 25. What is qualitative capacity of acidity?**

- Acidity of water is its qualitative capacity to react with a strong base to a designated pH.

**Q 26. What is an another name of methyl orange acidity?**

- Another name of methyl orange acidity is mineral acidity.

**Q 27. What is the pH of methyl orange acidity?**

- The pH of methyl orange acidity is pH 4.

**Q 28. Which is the recommended colour in methyl orange acidity?**

- The recommended colour in methyl orange acidity is blue colour as it has a sharper colour change at pH 3.7.

**Q 29. What is an another name of Phenolphthalein acidity?**

- Phenolphthalein acidity is also known as carbon dioxide acidity.

**Q 30. Why phenolphthalein acidity is known as carbon dioxide acidity?**

- Phenolphthalein acidity is known as carbon dioxide acidity because due to dissolution of atmospheric carbon dioxide in water and release of $CO_2$ during algal photosynthesis in water.

**Q 31. How carbon dioxide acidity can be measured?**

- Carbon dioxide acidity can be measured by titrating with standard sodium hydroxide solution using methyl orange and phenolphthalein indicator.

**Q 32. How mineral acidity is measured?**

- Mineral acidity is measured by titrating a sample to pH of about 3.7, i.e. methyl orange acidity.

**Q 33. How acidity affects water quality?**

- Acidity affects water quality with change in chemical properties of water, adverse effects on aquatic flora and fauna and reduction in dissolved oxygen.

**Q 34. What is the method of analysis of acidity?**

- The method of analysis of acidity is acid base titration.

**Q 35. Which reagents are used for acidity determination?**

- Reagents used for acidity determination are:
  - (a) Sodium hydroxide solution (0.02 *N*)
  - (b) Sodium carbonate
  - (c) Phenolphthalein indicator, and
  - (d) Methyl orange indicator.

**Q 36. What causes sluggish response in acidity by potentiometric titration?**

- In potentiometric titration oily matter, suspended solids, precipitation or other wastewater may coat the glass electrode and causes sluggish response in acidity.

**Q 37. What contributes to strong acidity?**

- Strong mineral acids such as $H_2SO_4$, $HNO_3$ and HCl contribute to strong acidity.

**Q 38. What are the contributing factors to weak acidity?**

- Carbonic and acetic acid contribute to weak acidity.

**Q 39. Which hydrolyzing salts contributes to acidity?**

- Hydrolyzing salts such as iron or aluminium sulphate may contribute to acidity depending on the method of determination.

**Q 40. In acidity how are resulting titrations identified?**

- Resulting titrations in acidity are identified traditionally as methyl orange acidity (pH 3.7) and phenolphthalein acidity (pH 8.3) regardless of actual method of measurement.

**Q 41. How is mineral acidity calculated?**

- Mineral acids contributing to mineral acidity can be calculated by titrating or neutralizing sample to pH 4.3 with standard alkali using methyl orange indicator to pH 3.7 using bromophenol blue.

**Q 42. In mineral acidity estimation, other than neutralize sample to pH 4.3, which other indicator and pH is used?**

- Other than neutralizing the sample to pH 4.3, the other indicator used is bromophenol blue to pH 3.7.

**Q 43. In water sample $CO_2$-bicarbonates -carbonates system corresponds to?**

- In water sample containing $CO_2$-bicarbonates-carbonates corresponds to stoichiometric neutralization of carbonic acid to bi-carbonate.

**Q 44. What are prime sources of interference in acidity?**

- Colour, turbidity, iron, aluminium and residual chlorine are prime sources of interference in acidity.

**Q 45. How can interference in acidity be avoided?**

- Interference due to colour and turbidity can be avoided using Potentiometric titration.

**Q 46. How can residual chlorine be removed?**

- Residual chlorine can be removed by adding Sodium thiosulfate.

**Q 47. Which types of processes influence chemical speciation in acidity?**

- Chemical and biological processes influence chemical speciation in acidity.

**Q 48. What is the normality of NaOH used in standardization process of acidity?**

- The normality of NaOH used in standardization process of acidity is 0.02 *N*, i.e. 1 mL of 0.02 *N* NaOH = 1 mg $CaCO_3$.

**Q 49. In case if normality of NaOH is other than 0.02 *N*, how can acidity be calculated?**

- In case if normality of NaOH is other than 0.02 *N* the acidity can be calculated: Acidity mineral as total of mg $CaCO_3$/L = A or B × *N* × 50 × 1000/ mL of sample

  *Where,*

  A = mL of NaOH required for sample pH 4.3-4.4
  B = mL of NaOH required for sample to pH 8.3
  N = Normality of NaOH used
  50 = Equivalent weight of $CaCO_3$, and
  1000 = Conversion factor to litre.

**Q 50. What is the effect of dissolved gases during water sampling?**

- Dissolved gases such as $CO_2$, $H_2S$ or $NH_3$ which may be lost or gained during sampling, storage or titration contribute to acidity of the water sample.

## 2.7 Alkalinity

**Q 1. What is alkalinity?**

- Alkalinity is the measure of substances in water that have acid-neutralizing property.

**Q 2. Is alkalinity a pollutant?**

- No, alkalinity is not a pollutant.

**Q 3. What does alkalinity measures?**

- Alkalinity measures strength of an acid or base.

**Q 4. What does alkalinity indicates in a solution?**

- Alkalinity in a solution indicates its power to react with acid.

**Q 5. In alkalinity, how pH can be illustrated in pure water?**

- In pure water, pH of exactly 7.0 contains no acid, no base and no alkalinity.

**Q 6. In alkalinity, how pH can be illustrated in buffered water.**

- In buffered water pH 6.0 has high alkalinity.

**Q 7. How does pH changes in pure water on addition of weak acid?**

- On addition of weak acid, pH of pure water will change instantly, it becomes acidic.

**Q 8. How does pH changes in buffered water on addition of weak acid?**

- In buffered water, pH doesn't changes easily because salts in buffer solution absorb acid and keep it from expressing itself.

**Q 9. How is alkalinity important for aquatic life?**

- Alkalinity is important for fish and aquatic life because it protects against pH change.

**Q 10. What are the main sources of alkalinity?**

- Main sources of alkalinity are rocks which contain carbonates, bi-carbonates and hydroxide compounds. Borates, silicates and phosphate may also contribute to alkalinity.

**Q 11. Why does Lime stone have good buffering capacity?**

- Limestone is rich in carbonates, so water flowing through lime stone regions has high alkalinity hence has good buffering capacity.

**Q 12. Why granite has poor buffering capacity?**

- Granite doesn't have minerals that contribute to alkalinity and are rich in minerals which have low alkalinity and hence have poor buffering capacity.

**Q 13. How does alkalinity occur?**

- Alkalinity commonly occurs due to dissolved salts of carbonates, bi-carbonates, borates, silicates, phosphates and hydroxyl ions.

**Q 14. Define alkalinity.**

- Alkalinity is defined as the capacity of an aqueous solution to neutralize acid contents.

**Q 15. At what pH alkalinity is determined as phenolphthalein alkalinity?**

- Alkalinity is determined as phenolphthalein alkalinity at pH 8.3.

**Q 16. At what pH alkalinity is determined as methyl orange alkalinity?**

- Alkalinity is determined as methyl orange alkalinity at pH 4.5.

**Q 17. Which titrant is used for determination of alkalinity?**

- The titrant used for determination of alkalinity is standard sulphuric acid (0.02 *N*).

**Q 18. How alkalinity is determined?**

- Alkalinity is determined as phenolphthalein alkalinity at pH 8.3 and total alkalinity with methyl orange at pH 4.5 by titrating against standard sulphuric acid (0.02 *N*).

**Q 19. What are interference causing agents in alkalinity?**

- Colour, turbidity, iron, aluminium and residual chlorine are interference causing agents in alkalinity.

**Q 20. How residual chlorine can be removed?**

- Residual chlorine can be removed by addition of calculated volume of sodium thiosulphate solution.

**Q 21. How colour and turbidity can be avoided in alkalinity analysis?**

- Colour and turbidity can be avoided in alkalinity analysis by titrating the sample potentiometrically.

**Q 22. What reagents are used for determination of alkalinity?**

- For determination of alkalinity the reagents used are:
  a) Standard $H_2SO_4$ solution (0.02 *N*)
  b) Solid $Na_2CO_3$
  c) Phenolphthalein indicator and
  d) Methyl orange indicator.

**Q 23. Which titrant should be used for standardization in alkalinity?**

- Sulphuric acid (0.02 *N*) titrant should be used for standardization in alkalinity.

**Q 24. What is the end point in standardization process of alkalinity?**

- Orange colour shows the end point in standardization process of alkalinity.

**Q 25. How 1 *N* HCl or 1 *N* $H_2SO_4$ can be calculated in standardization process in alkalinity?**

- 1 *N* HCl or 1 *N* $H_2SO_4$ = Wt. of $Na_2CO_3$ × 1000/53 × mL HCl or $H_2SO_4$ (eq. wt. of $Na_2CO_3$ = 53).

**Q 26. How phenolphthalein alkalinity is calculated?**

- Alkalinity (Phenolphthalein) as $CaCO_3$ can be calculated by, mg/L= mL (acid) × N (acid) × 1000/mL of sample taken, (Or) Alkalinity (Phenolphthalein) as $CaCO_3$ can be calculated by, mg/L = mL (acid) × N (acid) × 1000 × 50/ mL of sample taken, (1 mL HCl/$H_2SO_4$ = 50 mg $CaCO_3$).

**Q 27. How can total alkalinity be calculated?**

- Total alkalinity as $CaCO_3$ can be calculated by, mg/L = mL (acid for phenolphthalein) + mL (acid for methyl orange) × *N* (acid) × 1000/mL sample taken. (Or)

Total alkalinity as $CaCO_3$, mg/L = mL (acid for phenolphthalein) + mL (acid for methyl orange) × *N* (acid) x 50 × 1000/mL of sample taken (1 mL 1 *N* $HCl/H_2SO_4$ = 50 mg $CaCO_3$).

**Q 28. What is the relationship between different types of alkalinity?**

- Relationship of alkalinity can be expressed as,

$$P = 0, P < 1/2T, P = 1/2T, P > 1/2T, P = T;$$

where P = Phenolphthalein alkalinity and T = Total alkalinity (methyl orange).

**Q 29. How alkalinity is associated with hydroxyl, carbonate and bicarbonate ions?**

- Alkalinity is associated with hydroxyl, carbonate and bicarbonate ions as follows:

| Values of P & T | $OH^-$ | $CO_3^{2-}$ | $HCO_3^-$ |
|---|---|---|---|
| P = 0 | 0 | 0 | T |
| P < 1/2T | 0 | 2P | T-2P |
| P = 1/2T | 0 | 2P | 0 |
| P > 1/2T | 2P-T | 2(T-P) | 0 |
| P = T | T | 0 | 0 |

**Q 30. How is alkalinity expressed?**

- Alkalinity is expressed in terms of mg/L as $CaCO_3$.

**Q 31. How does alkalinity occur in natural water?**

- Major portion of alkalinity in natural water is caused by hydroxide, carbonates and bicarbonates.

**Q 32. Alkalinity is a measure of water–explain it.**

- Alkalinity of water is a measure of its capacity to neutralize acid. Alkalinity is a measure of water's ability to absorb $H^+$ ions without significant pH change i.e. alkalinity is a measure of buffering capacity of water. It is the sum of all titrable bases.

**Q 33. What is the significance of alkalinity due to other materials?**

- Alkalinity due to other materials in natural water is insignificant and may be ignored.

**Q 34. Which ions concentration is higher in alkalinity?**

- Bicarbonates, calcium and magnesium ions concentration is higher in alkalinity.

**Q 35. In natural water, how is alkalinity caused?**

- In natural water most of alkalinity is caused due to atmospheric carbon dioxide.

**Q 36. Which species are formed when free carbon dioxide gets dissolved into water?**

- When free carbon dioxide gets dissolved into water it forms carbonic acid ($H_2CO_3$) which further dissociates into $H^+$ and $HCO_3^-$, the $HCO_3^-$ further dissociates into $H^+$ and $CO_3^-$.

**Q 37. Is alkalinity harmful to Human being?**

- Alkalinity is not harmful to human being. Water supplies with less than 100 mg/L of alkalinity are desirable for domestic use.

**Q 38. How does alkaline water taste?**

- In large quantities, alkalinity imparts bitter taste to water.

**Q 39. What is the role of alkalinity in chemical coagulation process?**

- In chemical coagulation process alkalinity must be present in excess to neutralize acid released by coagulant.

**Q 40. Why does pH change due to aeration of water?**

- It is common practice to aerate water for removal of dissolved carbon dioxide. As carbon dioxide is acidic in nature, its removal tends to decrease pH (alkaline). Thus, high alkaline water tends to have high pH after aeration and low alkaline water tend to have low pH after aeration.

**Q 41. What is the Indian Standard Drinking Water Specification for alkalinity?**

- According to IS: 10500 (2012), 200 mg/L (Max.) is the standard of alkalinity in drinking water.

**Q 42. What happens if water has alkalinity above 200 mg/L?**

- Above 200 mg/L water becomes unpleasant to taste.

**Q 43. How is $H_2SO_4$ standardized in alkalinity?**

- In alkalinity $H_2SO_4$ is standardize against standard $Na_2CO_3$ (0.05 *N*).

**Q 44. What is the normality of sodium carbonate solution used in alkalinity?**

- The normality of sodium carbonate solution used in alkalinity is 0.05 *N*.

**Q 45. What is the pH of phenolphthalein indictor in alkalinity?**

- pH of phenolphthalein indicator used in alkalinity is pH 8.3.

**Q 46. What is de-colorization of phenolphthalein indicator indicates?**

- De-colorization of phenolphthalein indicator indicates complete neutralization of $OH^-$ ions and half of $CO_3$ ions.

**Q 47. What is the effect of dissolved gases during sampling in alkalinity?**

- Dissolved gases during sampling contributes to alkalinity, such as $CO_2$, $H_2S$ or ammonia which may be lost or gained during sampling, storage or titration.

## 2.8 Hardness

**Q 1. Define hardness.**

- Water hardness is the traditional measure of the capacity of water to precipitate soap.

**Q 2. What are the causes of hardness in water?**

- Hardness in water is not a specific constituent but is a variable and complex mixture of cations and anions. It is caused by dissolved polyvalent metallic ions.

**Q 3. What are the different components that participate in hardness?**

- In fresh water, the principle hardness causing ions are calcium and magnesium which precipitate soap. Other polyvalent cations also may precipitate soap, but often are in complex forms, frequently with organic constituents and their role in water hardness may be minimal and difficult to define.

**Q 4. What is total hardness?**

- Total hardness is defined as the sum of calcium and magnesium hardness. Total hardness = Calcium hardness + Magnesium hardness.

**Q 5. Hardness is expressed in which unit?**

- Hardness is expressed in milligram per litre (mg/L) as $CaCO_3$ or in ppm (parts per million).

**Q 6. What are the types of hardness?**

- In addition to total hardness it is classified in two ways:
  i) Hardness with respect to metallic ions and
  ii) Hardness with respect to anions associated with metallic ions.

**Q 7. What is the hardness with respect to metallic ions?**

- Hardness with respect to metallic ions is the hardness caused by calcium and magnesium metal ions in water.

**Q 8. How is drinking water classified on the basis of hardness?**

- The degree of hardness of drinking water has been classified in terms of equivalent $CaCO_3$ concentrations as: Soft: 0-60 mg/L, Medium: 60-120 mg/L, Hard: 120-180 mg/L and Very Hard: >180 mg/L.

**Q 9. What is temporary and permanent hardness?**

- Temporary hardness is the hardness which can be removed by boiling and permanent hardness is the one which can't be removed by boiling.

**Q 10. What is carbonate hardness or temporary hardness?**

- Carbonate hardness or temporary hardness refers to the amount of carbonate and bicarbonate in solution that can be removed or precipitated by boiling. This type of hardness is responsible for deposition of scale in hot water pipes and kettles. When total hardness is numerically greater than that of total alkalinity, it is called as "carbonate hardness".

**Q 11. What is non-carbonate hardness or permanent hardness?**

- The amount of hardness in excess of total alkalinity expressed as $CaCO_3$ is non-carbonate hardness. Non-carbonate hardness is caused by the association of hardness causing cations with sulphate, chloride or nitrate and is referred to as permanent hardness. This type of hardness cannot be removed by boiling.

**Q 12. What are the methods for determination of hardness?**

- Two methods are presently used as a standard methods for determination of hardness 1) Calculation method and 2) EDTA Titrimetric method.

**Q 13. Which calculation method is for determination of hardness?**

- The most accurate method of determining hardness is by a calculation based upon divalent ions found through a complete cations analysis. This method is to be preferred where complete analysis of water is available. The concentration of individual divalent cations of importance can also be determined separately by standard atomic absorption or inductively coupled plasma procedures. It has been shown that some hard waters contain appreciable amount of strontium. Thus, if strontium is not included in calculations, for total hardness, the calculated values may be in considerable error. Calculation of the hardness caused by each ion is performed by use of the general formula. Hardness (in mg/L) as = $m^{2+}$ (in mg × 50) where, $m^{2+}$ represent any divalent metallic ion.

**Q 14. What is EDTA titrimetric method?**

- When EDTA (as a titrant) is added into a water sample in presence of buffer solution, inhibitor along with EBT indicator, calcium ions are complexed with EDTA to form a wine red colour, which on titration resulted into blue colour as end point. In case of magnesium hardness, NaOH addition raised the pH to about 12 and murexide indicator resulted into pink colour which on titration gives purple colour as end point.

**Q 15. State the formula for calculation of hardness.**

- Hardness (mg/L) as $CaCO_3$ can be calculated Volume of EDTA required × 1000/mL of water sample taken.

**Q 16. How magnesium hardness in calculated?**

- Magnesium hardness is calculated as: Magnesium hardness = Total hardness – Calcium hardness.

**Q 17. Why buffer solution is used in hardness estimation?**

- Buffer solution is used for estimation of hardness as by its addition the pH gets raised upto 10.0 ± 0.01.

**Q 18. Why 2 *N* NaOH is added in estimation of calcium hardness?**

- Addition of 2 *N* NaOH during estimation of calcium hardness suppresses the magnesium ions in water sample and only calcium ions remains in solution.

**Q 19. Which indicators are used in hardness estimation?**

- For estimation of total hardness EBT (Eriochrome Black T) indicator is used; whereas, for calcium hardness Ammonium purporate (Murexide) is used.

**Q 20. What happens after addition of EBT indicator in water sample?**

- After addition of EBT indicator in water sample in presence of other reagents, EBT reacts with metal ions (divalent) and forms a wine red colour.

## 2.9 Chlorides

**Q 1. What is a chloride?**

- Chloride is a compound formed by the combination of chlorine with another element (e.g. sodium chloride, NaCl) or with a group of atoms (e.g. methyl chloride, $CH_3Cl$).

**Q 2. Usually chlorides are present in water as.**

- Chlorides are present in water usually as NaCl, $MgCl_2$ and $CaCl_2$.

**Q 3. Give classification of freshwater on the basis of chloride concentration.**

- On the basis of chloride concentration, fresh water is classified as; from 0.0 to 500.0 mg/L as chloride content, 501.0 to 1000.0 mg/L as salt water and above 1001.0 mg/L as brackish water.

**Q 4. What is the relation between chloride content and mineral content in a water sample?**

- The chloride content normally increases as mineral content increases in a water sample.

**Q 5. Which water source has more chloride concentration?**

- River, groundwater, sea and ocean water normally have more chloride concentration than other water sources.

**Q 6. What is the source of chloride in inland areas?**

- Salt sprays from oceans which carry with them droplets or minute salt crystals are the source of chloride in inland areas.

**Q 7. What is the average chloride discharge through urine per day per person?**

- The average chloride discharge through urine per day per person is 6 g of chlorides.

**Q 8. State significance of chloride estimation in a water sample.**

- At concentration above 250 mg/L, chloride gives salty taste to water which is objectionable to many people.

**Q 9. What are isochlors?**

- Locations in a map connected by lines with similar chloride concentrations are called as isochlors.

**Q 10. What are the effects of chlorides on plants?**

- Evaporation of water increases the chloride concentration at the roots of irrigated plants. This leads to difficulty for the crops to take up water due to osmotic pressure between the outside of the plants and within the plant cell.

**Q 11. For salt sensitive crops, what should be the chloride concentration in water?**

- For salt sensitive crops, chloride concentration in water should be at or below drinking water standard (250 mg/L).

**Q 12. Which procedure is employed for measurement of chlorides in water?**

- Chlorides can be measured by volumetric procedure.

**Q 13. Which methods can be adopted for estimation of chlorides?**

- Three methods can be adopted for estimation of chlorides a) Titration against standard mercuric nitrate solution b) Argentometic titration and c) Potentiometric titration.

**Q 14. Name the standard method used for estimation of chloride concentration.**

- Mohor's method is the standard method used for estimation of chloride concentration.

**Q 15. Which type of titration is involved in Mohor's method?**

- In Mohor's method, argentometric titration is involved.

**Q 16. Which titrant and indicator is used in Mohor's method?**

- In Mohor's method, 0.0141 *N* silver nitrate is used as a titrant and 1 mL of potassium chromate is used as an indicator.

**Q 17. Each one mL of 0.0141 *N* silver nitrate is equivalent to how many chloride ion?**

- Each one mL of 0.0141 *N* silver nitrate is equivalent to 0.5 mg of chloride ion.

**Q 18. Which standard solution is used to standardize Silver nitrate in Mohor's method?**

- Standard Sodium chloride solution (0.0141 *N*) is used to standardize Silver nitrate in Mohor's method.

**Q 19. Why silver nitrate solution (0.0141 *N*) should be stored in amber coloured volumetric flask?**

- Silver nitrate solution (0.0141 *N*) should be stored in amber coloured volumetric flask so as to prevent its oxidation due to its exposure to solar radiation.

**Q 20. During titration which reactions takes place?**

- During titration following reactions takes place:

$$Ag^+ + Cl^- \longrightarrow AgCl \qquad (1)$$

and $2Ag^+ + CrO_4^{2-} \longrightarrow AgCrO_4$ (2).

During titration, chloride ion are precipitated as white silver chloride (1). When potassium chromate is used as an indicator, after extinction of chloride ions, silver ions reacts with chromate ions and solubility product of silver chromate is exceeded and it begins to form reddish-brown precipitate (2).

**Q 21. Why indicator blank titration is carried out?**

- Distilled water may also contain trace quantity of chloride ions which was used for preparation of reagents. Its interference tends to be removed hence blank titration is carried out.

**Q 22. What is the volume of chromate indicator used in titration?**

- The volume of chromate indictor used in titration is 1 mL.

**Q 23. Which type of indictor potassium chromate is?**

- Potassium chromate is a metalochrome indicator.

**Q 24. What should be the potassium chromate indicator concentration during the titration?**

- The potassium chromate indicator concentration during the titration should be 0.005 to 0.0025 M.

**Q 25. Why estimation of chloride ion is carried out at pH range of 7-8?**

- Estimation of chloride ion should be carried out at pH range of 7-8 because $Ag^+$ is precipitated as AgOH at high pH levels and $Cr_2O_7^{2-}$ at low pH levels.

**Q 26. State formula used for computation of chloride ion concentration.**

- Chloride ion concentration can be calculated as:

  $Cl^-$ (mg/L) = (Sample-Blank) × Normality of $AgNO_3$ used × 35.45 × 1000/mL of water sample taken. Where, sample and blank means volume of the standerd Silver nitrate solution required in mL while titrating.

**Q 27. What is the Indian Standard Drinking Water Specification for chlorides according to IS 10500:2012?**

- The Indian Standard Drinking Water Specification for chlorides according to IS 10500:2012 is 250 mg/L (Max.).

**Q 28. How can interference causing agents be removed in chloride estimation?**

- Interference causing agents such as sulphate, thiosulphate and sulphite ion can be removed in chloride estimation by treatment of water sample with hydrogen peroxide.

**Q 29. What is the fate of $MgCl_2$ in boiler feed water?**

- In boiler feed water, $MgCl_2$ undergoes hydrolysis under intense temperature and pressure and leads to the formation of HCl which causes corrosion of boiler.

**Q 30. How wastewater samples colour and turbidity can be removed before chloride estimation?**

- Wastewater samples colour and turbidity can be removed by addition of alum and supernatant liquid before chloride estimation.

## 2.10 Fluoride

**Q 1. What is a fluoride?**

- Fluoride is the safest and most economical tool for dental caries prevention. Fluorides are compounds of the element "fluorine". The term fluorine is derived from a Latin word '*floure*' which means 'to flow'. Fluoride accounts for about 0.3 g/kg of the earth's crust and exists in the form of fluorides in a number of minerals of which fluorspar, cryolite and fluoraptite are the most common.

**Q 2. What is the oxidation state of fluoride?**

- The oxidation state of fluoride is -1.

**Q 3. What are the sources of fluoride?**

- Sources of fluoride are classified into following three ways: natural, anthropogenic and industrial. Fluoride and apatite are most commonly occuring minerals in different types of rocks and are the principal sources of fluorides in groundwater. Fluorides also occurs in lattices of amphibole, mica and garnet etc.

**Q 4. What are the natural sources of fluorides?**

- Fluorides are released into environment naturally through weathering and dissolution of minerals in emission from volcanoes and in marine aerosols. Estimate of the annual release of HF from volcanic sources through passive degassing and eruptions ranges from 60 to 6000 kilo tones of which, approximately 10% may be introduced directly into the stratosphere.

**Q 5. What are the anthropogenic sources of fluoride?**

- Fluorides are released into environment by coal combustion and process waters and waste from various industrial processes including steel manufacturing primary aluminium, copper, nickel production, phosphate ore processing, phosphate fertilizers production and use glass, brick and ceramic manufacturing and glue and adhesive production. The use of fluoride containing pesticides as well as controlled fluoridation of drinking water supplies also contributes in release of fluoride from anthropogenic sources.

**Q 6. What are the industrial sources of fluorides?**

- Hydrogen fluoride (HF) is an important industrial compound with estimated annual world consumption, of more than 1 million tonnes. It is one of the major sources of fluoride. It is used mainly in production of synthetic cryolite, aluminium fluoride, motor gasoline alkylates and chlorofluorocarbons. It is also used in etching semiconductor devices, cleaning and etching glass, cleaning brick and tanning leather. Hydrogen fluoride is also found in commercial rust removers. Based on the available data, phosphate ore production and use as well as aluminium manufacturers are the major industrial sources of fluoride released into the environment. The main sources of fluoride emissions are clay brick manufacturing, aluminium smelting and electricity production industries.

**Q 7. What are the Indian Standards for fluoride in drinking water?**

- Fluoride concentration limits (mg/L) in drinking water are as follows:

| Drinking water Standard | Minimum acceptable concentration (mg/L) | Maximum allowable concentration (mg/L) |
|---|---|---|
| Indian Standard (IS 10500:2012) | 1.0 | 1.5 |
| Ministry of Urban Development, India | 1.0 | 1.5 |

**Q 8. State the significance of fluoride in drinking water.**

- Fluoride ions have dual significance in water supplies. High concentration of fluoride causes dental fluorosis (disfigurement of teeth). At the same time, a concentration less than 0.8 mg/L results in dental caries. Hence, it is essential to maintain fluoride concentration between 1.0 to 1.5 mg/L in drinking water.

**Q 9. What are the benefits of fluoride in water?**

- About 96% of fluoride in body is found in bones and teeth. Fluoride is essential for normal mineralization of bones and formation of dental enamel. The principle source of fluoride to the physiology of man is water and food. Hence, fluoride in water is beneficial.

**Q 10. What are the adverse effects of fluoride?**

- Adverse effects of fluoride includes:
  1) Fluoride in small dosages has remarkable influence on dental system by inhibiting dental caries with higher dosages cause dental and skeletal fluorosis.
  2) Fluorosis affects teeth, joints and bones. Due to this disease; teeth get discoloured and weak, joints get swollen and deformed and bones may suffer morbid enlargements. It weakens and deforms the human skeletal structure gradually dragging the patient to painful death.
  3) Fluoride in drinking water gets rapidly, totally and passively absorbed by the intestinal mucosa.
  4) Severe and widespread gastrointestinal problems, namely anorexia pain in the stomach, intermittent diarrhoea, chronic constipation, gas

formation and bloated feeling in stomach caused due to fluoride contaminated water have been observed.

**Q 11. What are the biological effects of fluoride?**

- Biological effects of fluoride includes:
  1) The concentration of fluoride upto 0.002 mg/L in air causes injury to vegetation.
  2) The concentration of fluoride upto 2.0 mg/L or more in water causes mottled enamel.
  3) The concentration of fluoride upto 3.1 to 6.0 mg/L in water causes osteoporosis.
  4) The concentration of 50 mg/L fluoride in food or water causes thyroid changes.
  5) The concentration of 100 mg/L fluoride in food or water causes growth retardation.
  6) Kidney changes are also caused, if the concentration of fluoride is >125 mg/L in food or water and
  7) In acute doses i.e. 2.5-5 mg/L causes death.

**Q 12. In which state is fluoride present in the atmosphere?**

- Fluoride in atmosphere is present in the form of gas and as particulates.

**Q 13. What is fluoridation?**

- The process of adding fluoride in water is known as fluoridation.

**Q 14. What should be added to fluorinate water?**

- In order to fluorinate water, salts containing fluoride ion are added. Sodium fluoride solution, fluosilicic acid, $H_2SiF_6$ and hydrofluoric acid are added in water. Flural $AlFSO_4.H_2O$, ammonium fluorosilicate and other reagents are also used for fluorination of water.

**Q 15. What is defluorination of water?**

- The process of removing fluoride from water is known as defluorination.

**Q 16. Which chemicals are used for defluorination of water?**

- The chemicals like magnesium salt and calcium phosphate are used for defluorination of water.

**Q 17. Which methods are used for defluoridation?**

- Strongly basic anion exchange resins, special activated coals, magnesia sorbents and activated alumina etc. have been used in sorption methods of removing fluorine from water.

**Q 18. Give reactions of fluoride removal in water.**

- 1) Defluorination of water with magnesium salt is carried out by adding lime which increases pH of water to 10.2 to 10.3 under this conditions magnesium hydroxide is formed. $MgCl_2 + Ca\ (OH)_2 \rightarrow CaCl_2 + Mg(OH)_2$.

Magnesium hydroxide reacts with fluoride ion to form an insoluble basic oxyfluoride of magnesium. $Mg(OH)_2 + NaF \rightarrow Mg(OH)F \downarrow + NaOH$.

2) Calcium phosphate which binds fluoride into a sparingly soluble compound. Calcium phosphate should preliminary be treated with 10% NaOH solution during which calcium oxyphosphate is formed $2Ca_3(PO_4)_2 + 3NaOH \rightarrow Ca_5(OH)(PO_4)_3 + Ca(OH)_2 + Na_3PO_4$. This calcium salt binds the fluoride ion as: $Ca_5(OH)(PO_4)_3 + NaF \rightarrow Ca_5F(PO_4)_3 + NaOH$.

**Q 19. Which is the standard method for the analysis of fluoride?**

- Among many methods suggested for determination of fluoride ion in water, colorimetric method (SPANDS) and ion selective electrode method are the most satisfactory standard methods used for fluoride determination.

**Q 20. What is the principle of SPANDS method?**

- Under acidic conditions, fluoride reacts with zirconium SPANDS dye lake solution and the colour get bleached due to formation of a colourless complex anion ($ZrF_6$). As the amount of fluoride increases colour produced becomes progressively lighter. Since bleaching is a function of fluoridation, it is directly proportional to the concentration of fluoride.

**Q 21. State the reaction occuring in SPANDS method.**

- The reaction occuring in SPANDS method is:
  (Zr- SPANDS lake) + $6F^- \rightarrow$ SPANDS red bleached + $ZrF_6$.

**Q 22. State the full form of SPANDS.**

- The full form of SPANDS is Sodium 2- (parasulfophenylazo)-1, 8-dihydroxy-3,6-napthalene disulphonate and is also called as 4,5-dihydroxy -3-(parasulfophenylazo)-2, 7-naphthalenedisulfonic acid trisodium salt.

**Q 23. Which law is followed in SPANDS method?**

- In SPANDS method, Beer's law in a reverse manner is obeyed.

**Q 24. State Beer's law.**

- Beer's law state that "When a beam of monochromatic light is passed through a medium, the intensity of light decreases exponentially as the concentration of absorbing substance increases".

**Q 25. How fluorosis occurs in human body?**

- Fluorosis occurs in human body in localities where artesian waters are used for drinking purpose. Fluoride irritates and corrodes all body passages, they enter into the blood chain and then cause fluorosis.

**Q 26. At what concentration fluoride is present in natural water?**

- Surface water contains insignificant quantities of fluoride but sub soil waters from artesian and other wells and also from springs which are used for local water supply contains increased doses of fluorine. Its concentration in such water may be as high as 6-10 mg/L.

**Q 27. What is toxicological effect of fluoride in drinking water?**

- Toxicological effect due to excess of fluoride in drinking water is bone cancer in young individuals.

**Q 28. Which factors affects fluoride content in groundwater?**

- Factors affecting fluoride content in groundwater includes:
  1) Size and distribution of fluoride and other fluoride bearing minerals in rocks and soils.
  2) Availability and accessibility of groundwater in contact with those minerals. Water sources near granite rocks containing more than 2.5 mg/L.
  3) Complexion of fluoride ion with aluminium, beryllium, ferric ion and series of mixed fluoride hydroxide complex with boron.
  4) High fluoride concentration is caused by excessive alkalinity and low calcium contents of soil and water. There is a negative correlation between fluoride and total hardness.

**Q 29. Which rocks and minerals are responsible for fluoride concentration in groundwater?**

- Fluoride minerals $Ca_5F(PO_4)_3$, fluorite and apatite minerals and granites shales and clays, sandstone, dolomite or dolomitic limestone are responsible for fluoride concentration in groundwater.

**Q 30. What are the effluent discharge standard for fluoride?**

- Effluent discharge standard by MoEF (1993) for fluoride is: Guideline value of 2 mg/L is set for discharge into inland surface water and 15 mg/L for discharge into public severs and marine coastal areas.

## ■ 2.11 Sodium and Potassium

**Q 1. What is the percentage of potassium available in plants?**

- The percentage of potassium available in plants is 1-4%.

**Q 2. What is the melting point of potassium?**

- The melting point of potassium is 63.65°C.

**Q 3. In India, in which soil potassium deficiency occurs?**

- Potassium deficiency occurs in acidic soil in India.

**Q 4. What are commercial sources of potassium?**

- Commercial sources of potassium include sulphate contain 48 to 50% $K_2O$ and muriate of potash (KCl) contain 60 to 63% $K_2O$.

**Q 5. What is the percentage of potassium present in the earth crust?**

- In the earth curst percentage of potassium present is 2 to 2.5%.

**Q 6. In which century was potassium discovered?**

- Potassium was discovered in $19^{th}$ century.

**Q 7. Why is sodium stored under kerosene?**

- Sodium is highly reactive in nature and hence it is stored under kerosene.

**Q 8. What percentage of potassium is present in water?**

- About 2 to 3 ppm of potassium is present in water.

**Q 9. What is the percentage of potassium present is fertilizer?**

- About 95% of potassium is present in fertilizer.

**Q 10. What is the importance of potassium in plant?**

- Potassium plays an important role in plant-water relationship by regulating opening of stomata and transpiration.

**Q 11. What is the atomic number of potassium?**

- The atomic number of potassium is 19.

**Q 12. What are the effects of salinity on soil's physical properties?**

- Soil water salinity affects soil's physical properties by causing fine particles to bind together into aggregates. This process is known as flocculation and is beneficial for soil aeration, root penetration and root growth.

**Q 13. Which nutrients are absorbed in maximum amount in plants?**

- Potassium is the major nutrient absorbed by plants followed by sodium and calcium.

**Q 14. What percentage of potassium content is in the earth crust?**

- Potassium content of earth crust is 2.4%.

**Q 15. How does potassium react with water?**

- The reaction between potassium and water is:

$$2K_{(s)} + H_2O_{(l)} \longrightarrow 2\ KO_{(aq)} + H_{2(g)}.$$

**Q 16. Which method is used for the estimation of sodium?**

- Sodium is estimated by emission spectroscopy or by flame photometer.

**Q 17. Where sodium is present ?**

- Sodium is present in number of minerals, the principal one being rock salt (Sodium chloride).

**Q 18. Why does in tropics soil potassium content low?**

- The amount of potassium in tropical soil may be low because of the origin of soil, high temperature and high rainfall in such areas.

**Q 19. In which condition nitrogen and phosphorus deficiency is observed?**

- Nitrogen and phosphorus deficiency occurs in leaching and fixation condition.

**Q 20. State the source of potassium in soil.**

- Soil potassium is derived by the decomposition of potassium containing primary minerals.

**Q 21. What are the primary sources of potassium available to plants?**

- Primary sources of potassium available to plants includes: potash feldspar, muscovite and biotite.

  Their availability to plants is in the order of biotite > muscovite > potash feldspar.

**Q 22. What are the secondary sources of potassium?**

- Secondary sources of potassium includes clay mineral, hydrous mica, vermiculites, chlorites, etc.

**Q 23. What is non-exchangeable potassium?**

- Non-exchangeable potassium is that potassium, which forms a part of crystal structure of un-weathered or slightly weathered secondary minerals.

**Q 24. What is the luxury consumption of potassium?**

- Potassium is one nutrient which exhibits the phenomenon of luxury consumption. If potassium is supplied in amounts more that what is needed for maximum crop yield, crops will utilize more potassium than what is required. This is called luxury consumption of potassium.

**Q 25. Which potassium containing minerals are highly resistant to weathering?**

- Potassium containing minerals such as feldspar and mica are highly resistant to weathering.

**Q 26. What is 'unavailable potassium' in soil?**

- Unavailable potassium is the primary mineral potassium (Feldspar and mica), 90-98% of total potassium in soil.

**Q 27. What is 'slowly available' potassium?**

- Slowly available potassium is secondary mineral potassium or reserve potassium.

**Q 28. What is 'readily available' potassium in soil?**

- Readily available potassium in soil is exchangeable potassium and soil solution potassium i.e. 0.1-2% potassium in soil.

**Q 29. What is the Indian Standard Drinking Water Specification for sodium (IS 10500:2012)?**

- No standard is recommended for sodium in Indian Standard Drinking Water Specification (IS 10500:2012).

**Q 30. What is the concentration of potassium in rain water?**

- Potassium concentration in rain water is generally low, with an average annual concentration of 0.1 mg/L.

**Q 31. Name the methods used for potassium determination.**

- Potassium can be determined by gravimetric, titrimetric, absorption emission and spectrophotometric method.

**Q 32. What is the importance of potassium?**

- Potassium is a natural and essential element in plant and animal growth.

**Q 33. Which metals affects the activity of potassium?**

- Calcium and sodium affects the activity of potassium.

**Q 34. What is the role of potassium in crop growth?**

- Potassium deficiency reduces crop yield.

**Q 35. What is the symbol of potassium?**

- The symbol of potassium is K, i.e. Kalium, the medieval latin for potash, which may have been derived from the arabic word *qali,* meaning alkali.

**Q 36. What is the atomic number of sodium?**

- The atomic number of sodium is 11.

**Q 37. Is sodium found in drinking water?**

- Yes, sodium is found in low levels in drinking water.

**Q 38. What is the atomic mass of sodium?**

- The atomic mass of sodium is 22.98 atomic mass unit.

**Q 39. What does SAR stands for?**

- SAR stands for sodium absorption ratio.

**Q 40. Why is sandy soil not affected by sodium?**

- Sandy soil is not affected by sodium because of its low clay content.

**Q 41. Why do sodium ions destabilize soil aggregates?**

- Sodium ion breaks up clay mineral in soil. Sodium ion is relatively large and disrupt the pores that hide particles of clay together. This causes clay platelets to separate when this separation occurs. The clay particles expands causing swelling and soil dispersion.

**Q 42. Where is sodium bromide found in nature?**

- The rich source of sodium bromide in nature is the Dead Sea.

**Q 43. Why is sodium found in compound form in nature?**

- Sodium is a highly reactive element when it is in pure form, it reacts with elements and moisture in air almost immediately. It is more stable in compound form thus, it is so found in nature.

**Q 44. Which kind of metal is sodium?**

- Sodium is an alkali metal.

**Q 45. What is the melting point of sodium?**

- The melting point of sodium is 370.87 K or 97.72°C or 207.9° F.

**Q 46. What is the boiling point of sodium?**

- The boiling point of sodium is 1156 K or 883°C or 1621° F.

**Q 47. Are some soils more sensitive than other to saline or sodic soil?**

- Yes, irrigation water is suitable for one soil may not be suitable for another. Use of saline and sodic water for irrigation can cause harm to soils, predominate by silt or clay.

## 2.12 Sulphate

**Q 1. What is the sulphate?**

- Sulphate is a naturally occurring substance that contains sulphur and oxygen.

**Q 2. How does sulphate gets into water?**

- Sulphate may be bleached from soil and is commonly found in most water supplies.

**Q 3. By which method sulphate is estimated?**

- Sulphate is estimated by Barium chloride method.

**Q 4. What is the role of conditioning reagent in Barium chloride method?**

- Conditioning reagent provides acidic condition for estimation of sulphate in Barium chloride method.

**Q 5. Why $BaCl_2$ crystals are used in estimation of sulphate?**

- Barium chloride crystals are used in estimation of sulphate because sulphate ion ($SO_4$) is precipitated with $BaCl_2$ so as to form $BaSO_4$. [$SO_4^- + BaCl_2 \rightarrow BaSO_4 + 2Cl^-$].

**Q 6. What are the sources of sulphate in water?**

- Most of the sources of sulphate in water originate from oxidation of sulphite ore. The solution of gypsum and anhydrite atmospheric sulphur dioxide is formed by the combustion of fossil fuels. Sulphur trioxide ($SO_3$) produced by photolytic oxidation of sulphur dioxide which reacts with water vapour to form sulphuric acid which is precipitated as acid rain or snow.

**Q 7. What are the effects of sulphate on human heath?**

- Sulphate is generally considered as non-toxic. The consumption of drinking water containing high amount of magnesium and sodium sulphate may result in intestinal discomfort, diarrhoea and dehydration.

**Q 8. What are the interferences in sulphate estimation?**

- Colour, turbidity and silica in the concentration of 500 ppm interferes in the estimation of sulphate.

**Q 9. What are the problems caused by sulphate in water supplies?**

- Sulphate causes scaling in water supplies and the problem of odour and corrosion in wastewater treatment, due to its reduction to $H_2S$.

**Q 10. What is the effect of organic matter on sulphate estimation?**

- In the presence of organic matter, precipitation of barium sulphate may not be satisfactory.

**Q 11. How is sulphate removed from drinking water?**

- Sulphate can be removed from drinking water by demineralization through the use of reverse osmosis or distillation.

**Q 12. What is the guideline value of sulphate?**

- The guideline value of sulphate is less than 500 mg/L based on inoffensive taste and potential for laxative effect.

**Q 13. How is sulphate removed from municipal water supplies?**

- Ion-exchange can be used for sulphate removal from municipal water supplies.

**Q 14. What is the standard sulphate concentration in water from aesthetic point of view?**

- According WHO guideline, 400 mg/L of sulphate is the standard value in water, from aesthetic point of view.

**Q 15. State the environmental significance of sulphate.**

- Environmental significance of sulphate includes:
  1) Water containing appreciable quantity of sulphate forms hard scales in boilers and heat exchangers.
  2) Sulphate also causes odour and corrosion of sewer in anaerobic conditions as it gets converted into hydrogen sulphide.

**Q 16. What is the impact of sulphate in industries?**

- Sulphate forms hard scale in boilers and heat exchanger in industries.

**Q 17. What is the role of buffer solution in sulphate estimation?**

- Barium sulphate crystal formation gets enhanced in the presence of buffer solution in sulphate estimation.

**Q 18. What precaution can be taken to remove interferences in sulphate estimation?**

- Filtration can remove interferences of colour and turbidity in sulphate estimation.

**Q 19. What is the range of turbidometric method of sulphate estimation?**

- Turbidometric method of sulphate estimation is in the range of 1-40 mg $SO_4^{2-}$/L.

**Q 20. What is the role of magnetic stirrer in sulphate estimation?**

- Magnetic stirrer is used in sulphate estimation to get a constant stirring speed, as it will lead to the uniform formation of barium sulphate crystals.

**Q 21. What are the effects of sulphate on human beings?**

- Sulphate beyond 200 ppm causes gastrointestinal irritation when magnesium and sodium are present.

**Q 22. Why it is important to estimate sulphate in public water supplies?**

- Sulphate estimation is important in public water supplies because of its catalytic effects on humans when present in excessive amount.

**Q 23. What is the standard guideline of sulphate in effluent discharge?**

- According to MoEF (1993), 1000 mg/L (Max.) of sulphate is the standard value for discharge in surface water, public sewer and water for irrigation.

**Q 24. Which reaction occurs during estimation of sulphate?**

- $BaCl_{2(excess)} + SO_4^{2-} \longrightarrow BaSO_{4(ppt)}$ is the reaction occuring during the estimation of sulphate.

**Q 25. Why is acidic condition maintained during estimation of sulphate?**

- Acidic condition is maintained during estimation of sulphate so as to eliminate the possibility of precipitation of $BaCO_3$ which might occur in highly acidic conditions.

**Q 26. Which formula is used to estimate sulphate concentration?**

- The formula used in sulphate estimation is:
  $SO_4^{2-}$ mg/L = mg $SO^{2-}{}_4 \times 1000$/mL of sample.

**Q 27. Why is standard sulphate concentration taken upto 40 mg/L range?**

- Standard sulphate concentration is taken upto 40 mg/L because above this point, its accuracy decreases and $BaSO_4$ suspension losses its stability.

**Q 28. What is the concentration of standard sulphate solution?**

- The concentration of standard sulphate solution is 1 mL = 100 μg of sulphate.

**Q 29. Why is excess of $BaCl_2$ added in sulphate estimation?**

- Excess of $BaCl_2$ is added in sulphate estimation so as to produce sufficient common ions to precipitate sulphate.

**Q 30. Why is gravimetric method not used to estimation sulphate?**

- Gravimetric determination of sulphate is subject to a number of errors. Interference that leads to high results are suspended matter and silica hence, gravimetric method are not used to estimate sulphate.

## 2.13 Phosphate

**Q 1. What is phosphorous?**

- Phosphorous is a chemical element with symbol P. It exists in two forms: white phosphorous and red phosphorous. Phosphorous is an essential element for plants, though phosphorous content in plants is low.

**Q 2. Which phosphorous species are utilized by plants?**

- At soil pH, $H_2PO_4^-$ (dihydrogen phosphate) and $HPO_4^{2-}$ (monohydrogen phosphate) are the phosphorous species utilized by plants.

**Q 3. What is the reaction of phosphorous in the soil?**

- In alkaline soil, following reaction occurs with $CaCO_3$ whereby hydroxypatite is applied:
  $6HPO_4^{2-} + 10CaO_3 + 4H_2O \longrightarrow Ca_{10}(PO_4)_6(OH)_2 \downarrow + 10HCO_3^- + 20H^-$.

**Q 4. What happen when phosphorous concentration is above the permissible limit?**

- At higher concentration phosphate leads to eutrophication of water bodies.

**Q 5. What are the sources of phosphorous?**

- Sources of phosphorous includes fertilizer, sewage waste, sediments, pesticides and agriculture runoff.

**Q 6. What is eutrophication?**

- Eutrophication is defined as the state of water body which is manifested by an intense proliferation of algae and other higher aquatic plants.

**Q 7. Which two factors are responsible for eutrophication of a water body?**

- Nitrogen and phosphorous are responsible for eutrophication of a water body.

**Q 8. In which range phosphate is present in water?**

- Phosphate is present in water at a range of 0.1-2.8 mg/L.

**Q 9. What are the effects of phosphate?**

- Higher level of phosphate causes eutrophication of water bodies. In drinking water it also causes kidney stone.

**Q 10. Which are the standard methods for phosphate estimation?**

- The standard methods for phosphorous estimation includes:
  Vanodomolybdo phosphoric acid colorimetric method, Stannous chloride method, Ascorbic acid method and Automated ascorbic acid reduction method.

**Q 11. Which reagents are used in phosphate estimation in Stannous chloride method?**

- Reagents used in phosphate estimation in Stannous chloride method includes: Stock phosphate solution, phosphate working solution, ammonium molybdate and stannous chloride solution.

**Q 12. What is the principle involved in Stannous chloride method?**

- In acidic condition, orthophosphate reacts with ammonium molybdate to form molybdophosphoric acid. It reduced by stannous chloride to blue coloured complex. The intensity of blue colour is measured which is directly proportional to the concentration of phosphate present in the sample.

**Q 13. What is the significance of phosphate analysis?**

- The presence of phosphate in large quantities indicates pollution of fresh water. Though it poses problem in surface water its presence is necessary for biological degradation of wastewater.

**Q 14. At which wavelength phosphate estimation is carried out on spectrophotometer?**

- Phosphate estimation is carried out at 690 nm on spectrophotometer.

**Q 15. In which form phosphorous occurs in water and wastewater?**

- Phosphorous occurs in water and wastewater almost solely as phosphate.

**Q 16. Give the classification of phosphate.**

- Phosphate is classified as :

  (a) Orthophosphate, (b) Condensed phosphate and (c) Organically bound phosphate.

**Q 17. In which form phosphorous occurs in a water sample?**

- Phosphorus occurs in a water sample in particulates or detritus or in the body of aquatic organisms.

**Q 18. Which reaction takes place in stannous chloride method?**

- The reaction taking place in stannous chloride method is:
  $PO_4^{3-} + 12\ (NH_4)_2MoO_4 + 24H^+ \longrightarrow (NH_4)_3PO_4{\cdot}12MoO_3 + 21NH_4 + 12H_2O$.

**Q 19. Stannous chloride method is used for estimation of which form of phosphate?**

- Stannous chloride method is used for estimation of inorganic form of phosphate.

**Q 20. State the environmental significance of phosphate analysis.**

- The presence of phosphate in large quantities in fresh water indicates pollution through sewage and industrial wastes. Presence of phosphate is necessary for biological degradation of wastewater. Phosphate can be classified as total phosphate, inorganic phosphate and organic phosphate. Only inorganic compounds of phosphorous are of significant environmental importance.

**Q 21. Which form of phosphate is analyzed by colorimetrically?**

- Inorganic form of phosphate is analyzed colorimetrically.

**Q 22. What is the purpose of addition of $SnCl_2$ during colorimetric analysis of phosphate?**

- The purpose of addition of $SnCl_2$ during colorimetric analysis of phosphate is that in acidic condition, orthophosphate reacts with ammonium molybdate to form molybdophosphoric acid. It is reduced by stannous chloride to a blue colour complex.

**Q 23. Why acid digestion is needed for estimation of total phosphate?**

- All forms of phosphate, whether dissolved or particulate are converted into inorganic form (Phosphate) after acid digestion of sample. Therefore, for estimation of total phosphate acid, digestion is needed.

**Q 24. How do particulate phosphate and organic phosphate are calculated?**

- Particulate phosphate can be calculated as a difference between total phosphate concentration in unfiltered and filtered samples. Particulate phosphate = Total phosphate in unfiltered sample – Total phosphate in filtered samples.

  Organic phosphate can be calculated as difference between total phosphate and inorganic phosphate of sample. Organic phosphate = Total phosphate – Inorganic phosphate.

**Q 25. What is the recommended dietary allowance of phosphate in adults?**

- Recommended dietary allowance of phosphorous for adults is 800 mg/day.

**Q 26. Why does groundwater have higher phosphate concentration?**

- Water supplies may contain phosphate derived from natural contact with minerals or through contamination from application of fertilizer, sewage and industrial water. Groundwater therefore, is more likely to have higher phosphate concentration.

**Q 27. What is the range of phosphate in sewage?**

- In sewage concentration of phosphate (as total P) has been reported in the range of 9 mg/L to 15 mg/L.

**Q 28. What are the standards of phosphate in drinking water?**

- IS: 10500 (2012) = No standard

  WHO (1984) = No regulation or guidelines for phosphate.

  USEPA (1974) = Phosphate has not been regulated not even in secondary drinking water standard.

**Q 29. Which reagents are used for estimation of total phosphate?**

- Reagents used for estimation of total phosphate includes:
  1) Conc. Sulphuric acid
  2) Phenolphthalein indicator
  3) Conc. Nitric acid
  4) Sodium hydroxide 1 *N*, and
  5) Stock Phosphate solution.

**Q 30. Which chemicals are used for digestion of a sample?**

- Chemicals used for digestion of a sample includes perchloric acid, $H_2SO_4$-$K_2SO_4$, $H_2SO_4$-$HNO_3$ etc.

**Q 31. Which form of phosphate has environmental significance?**

- Inorganic phosphate is of environmental significance.

**Q 32. At which level phosphorous causes eutrophication problem?**

- Phosphorous causes eutrophication problem at near 0.05 mg/L.

**Q 33. What is the effluent discharge standard for phosphate?**

- Effluent discharge standard by MoEF (1993) for dissolved phosphate (as P) is 5 mg/L (Max.) for discharge inland and surface water bodies.

## 2.14 Nitrogen (Kjeldahl)

**Q 1. What is the Total Kjeldahl Nitrogen?**

- Total Kjeldahl Nitrogen (TKN) is the sum of organic nitrogen, ammonia ($NH_3$) and ammonium ($NH_4^+$) in the chemical analysis of soil, water or wastewater.

**Q 2. Who developed the Total Kjeldahl nitrogen estimation method?**

- Total Kjeldahl nitrogen estimation method was developed by Danish chemist Johan Kjeldahl in 1883.

**Q 3. What does the presence of total nitrogen indicate?**

- The presence of total nitrogen indicates a chemical evidence of recent organic pollution particularly of animal origin in water body.

**Q 4. What are the effects of enrichment of nitrogen?**

- Enrichment of nitrogen causes algal bloom, eutrophication of water bodies which inturn has adverse effects on fish and other organism and also on the water quality. Health disorder by nitrates leads to formation of methemoglobinemia (MetHb).

**Q 5. Why is the apparatus known as Kjeldahl apparatus?**

- This method was developed by Johan Kjeldahl and thus the apparatus is known as Kjeldahl apparatus.

**Q 6. What is the Indian Standard Drinking Water Specification for TKN (as nitrogen)?**

- According to Indian Standard Drinking Water Specification IS: 10500: 2012 for TKN (as N) no standard is stated.

**Q 7. What are the standards of TKN (as N) for effluent discharge?**

- A standard of TKN for effluent discharge according to MoEF (1993) is 100 mg/L maximum into inland surface water and marine coastal areas.

**Q 8. What is the role of nitrogen in plant growth?**

- Nitrogen serves as a nutrient for plant and it is essential for their growth.

**Q 9. Which chemical is used to preserve water sample for TKN estimation? State its quantity required.**

- For preservation of water sample for TKN estimation, 2 mL of $H_2SO_4$ per litre is added and stored at 4°C.

**Q 10. In which form nitrogen is generally present in water and wastewater?**

- Nitrogen is generally present in nitrate nitrogen, nitrite nitrogen, ammoniacal nitrogen and organic nitrogen, in water and wastewater.

**Q 11. Why TKN should be analyzed as soon as possible in the laboratory?**

- TKN should be analyzed as soon as possible because organic nitrogen may convert into ammonia in sample between time period of its collection and analysis.

**Q 12. What is the purpose of addition of conc. $H_2SO_4$ in digestion catalyst mixture?**

- The addition of conc. $H_2SO_4$ in digestion catalyst mixture is to catalyst organically bound nitrogen to convert it into ammonium sulphate. Potassium sulphate is added to raise the boiling point of sulphuric acid from 345°C to 370°C.

**Q 13. What should be the temperature during digestion of a sample for TKN estimation?**

- Temperature should be kept below 382°C during digestion of a water sample for TKN estimation.

**Q 14. Which two processes are involved in organic nitrogen determination?**

- In organic nitrogen determination, digestion and distillation processes are involved.

**Q 15. Which reagent is used for the digestion of sample?**

- Digestion reagent is used for digestion of sample.

**Q 16. What is the concentration of organic nitrogen in effluent?**

- Organic nitrogen found in effluent is as follows:
  1. For strong sewage TKN = 85 mg/L (Organic 35 mg/L + $NH_3^-N$, 50 mg/L);
  2. Medium strength TKN = 40 mg/L (Organic 15 mg/L + $NH_3^-N$, 25 mg/L) and
  3. Low strength TKN = 20 mg/L (Organic strength 8 mg/L + $NH_3^-N$, 12 mg/L).

**Q 17. What should be the pH of final solution before distillation?**

- The pH of final solution before distillation should be between 8 to 11.

**Q 18. Which reagent is used to raise pH of solution?**

- Sodium hydroxide–sodium thiosulphate is used to raise pH of solution.

**Q 19. What is the formula to calculate TKN?**

- The formula to calculate TKN (mg/L) = (A-B) × 280/mL of sample. Where, A = Volume of $H_2SO_4$ required for sample, mL and B = Volume of $H_2SO_4$ required for blank, mL.

**Q 20. If normality of $H_2SO_4$ is not 0.02 *N* then what formula should be used?**

- If normality of $H_2SO_4$ used is not 0.02 *N* then the formula used is:

  TKN (mg/L) = (A-B) × *N* × 14 × 1000 / mL of sample.

  Where, *N* = Normality of $H_2SO_4$ use as titrant.

**Q 21. Which titrant is used in TKN determination?**

- Standard $H_2SO_4$ solution of 0.02 *N* is used as a titrant in TKN determination.

**Q 22. Which indicator is use in TKN determination?**

- Mixed indicator (Methyl red and methylene blue) is used in TKN determination.

**Q 23. What is the end point of titration in TKN estimation?**

- Green colour changes to pink colour indicates end point of titration in TKN estimation.

**Q 24. During distillation process, boric acid's colour is converted into which colour?**

- During distillation process, boric acid's pink colour changes to green colour.

**Q 25. Which gases are evolved during digestion of a sample in TKN estimation?**

- Sulphur dioxide and carbon dioxide gases are evolved during digestion of a sample for TKN estimation.

**Q 26. In which form, organic nitrogen is present in nature?**

- Organic nitrogen is present in ionic form in nature.

**Q 27. When it is said to be Kjeldahl nitrogen?**

- If ammonia nitrogen is not removed in initial phase of the procedure, then it is said to be Kjeldahl nitrogen.

**Q 28. Why does the pink colour changes to green in TKN determination?**

- When ammonia is liberated during distillation process it gets absorb into boric acid and $H_3BO_3$ changes to $NH_4$-$H_2BO_3$, thus colour changes from pink to green.

**Q 29. What will happen when boric acid solution is titrated with standard 0.02 *N* $H_2SO_4$ solution?**

- When boric acid solution is titrated with standard 0.02 *N* $H_2SO_4$ solution, till the indicator turns a pale lavender colour, it indicates the ammonia present in the sample.

**Q 30. Which apparatus is used for determination of organic nitrogen?**

- Kjeldahl apparatus is used for determination of organic nitrogen.

**Q 31. Why distilled water is added after digestion?**

- Distilled water is added after digestion to make solution dilute.

**Q 32. When is ammonia liberated from the sample?**

- When the sample is kept for distillation, $NH_4H_2SO_4$ reacts with NaOH and releases $NH_3$.

**Q 33. Which method is used to determine ammonia or absorbed ammonia?**

- Ammonia or absorbed ammonia is determined either by titration or direct nesslerization.

**Q 34. What is the capacity of Kjeldahl flask?**

- The capacity of Kjeldahl flask is of 500 mL and 800 mL.

## 2.15 Nitrate

**Q 1. What is a nitrate?**

- Nitrate is a highly oxidized form of nitrogen compounds commonly present in natural waters.

**Q 2. What is the oxidation number of nitrogen in nitrate ion?**

- The oxidation number of nitrogen in nitrate ion is +5.

**Q 3. What are the sources of nitrate?**

- There are two sources of nitrate: natural sources and anthropogenic sources.

**Q 4. What are the natural sources of nitrate?**

- Natural sources of nitrate includes weathering, atmospheric washout, human and animal waste, atmospheric nitrogen, etc.

**Q 5. What are the anthropogenic sources of nitrate?**

- Anthropogenic sources of nitrate are chemical fertilizers, decay vegetable and animal matter, domestic effluents, sewage sludge disposal to land, industrial discharge and leachates from refuge dumps and agricultural practices.

**Q 6. How can nitrate contaminate water?**

- When nitrogen fertilizers are used to enrich soil, nitrates get entered into surface water and groundwater. Depending upon the situation, these sources can contaminate stream, river, lakes, groundwater especially wells. Although any well shallow, poorly constructed or improperly located well is more susceptible to nitrate contamination.

**Q 7. What are the standards for nitrate in drinking water?**

- Nitrate permissible limit concentration (mg/L) in drinking water is as follows :

| Drinking water standard | Permissible limit |
|---|---|
| USEPA standard | 10 mg/L |
| BIS (Indian Standard) | 45 mg/L |
| WHO standard | 10 mg/L |
| EU standard | 50 mg/L |

**Q 8. What is the significance of nitrate in drinking water?**

- Nitrate can affect red blood cells and reduce their ability to carry oxygen into body. In most adults and children these affect blood cells rapidly return to normal, but the infants under six months may develop a serious health condition due to lack of oxygen i.e. methemoglobinemia.

**Q 9. What happens if nitrate level is above the permissible limit?**

- Nitrate level at or above the permissible limit have been known to cause a potentially fatal blood disorder called as methemoglobinemia.

**Q 10. What is methemoglobinemia?**

- Methemoglobinemia or blue baby syndrome is a disease in which there is reduction in oxygen carrying capacity of blood due to combination of nitrate with blood haemoglobin.

**Q 11. What are the sources of nitrate in groundwater?**

- The sources of nitrate in groundwater include natural sources, waste material, raw crop agriculture and irrigated agriculture.

**Q 12. In which state nitrate is present in environment?**

- The nitrate is present in gaseous form such as $N_2$, $N_2O$ and $NO_2$. Nitrate that leaves the atmosphere can be converted back into elemental nitrogen.

**Q 13. What are the biological effects of nitrate?**

- Biological effects of nitrate above the drinking water standard causes blue baby syndrome in human infants. N-Nitrosamine can also lead to cancer.

**Q 14. Which wells are mainly contaminated by nitrate?**

- The wells which are poorly constructed or improperly located are more susceptible to nitrate contamination.

**Q 15. What does elevated nitrate level indicates?**

- Elevated nitrate level indicates possible presence of other contaminants such as disease causing and organic compounds that could cause health problems.

**Q 16. What are the significant sources of nitrate?**

- Significant sources of nitrate includes chemical fertilizers, decay vegetable, animal matter, domestic effluents, sewage sludge disposal to land, industrial discharge and leachates from refuse dumps and atmospheric wash out.

**Q 17. What is nitrification?**

- The process of addition of nitrate is known as nitrification.

**Q 18. What is the denitrification?**

- The process of the removal of nitrate is known as denitrification.

**Q 19. What are the different methods for nitrate removal?**

- Different methods for nitrate removal are: distillation, reverse osmosis, ion exchange, absorption charcoal and electro dialysis.

**Q 20. Can nitrate be removed by boiling?**

- No, nitrate can't be removed by boiling.

**Q 21. Why can nitrate not be removed by boiling?**

- Because nitrate is present in a stable form, it cannot be removed by boiling.

**Q 22. Can nitrate be removed by filtration?**

- No, nitrate can't be removed by filtration.

**Q 23. Why can nitrate not be removed by filtration?**

- Nitrate can't bind to any particle. It is present in stable form so by filtration nitrate can't be removed.

**Q 24. What are the standard methods for nitrate estimation?**

- Standard methods for nitrate estimation includes: UV-Visible spectrophotometer method, PDA method and Brucine method.

**Q 25. What is the principle of UV-Visible spectrophotometric method for determination of nitrate?**

- UV-Visible technique measures the absorbance of nitrate at 220 nm, which is suitable for screening uncontaminated water (low in organic matter). A second measurement made at 275 nm may be used to correct the nitrate value (because 275 nm is not absorbed by nitrate but absorbed by other matter). It follows Beer's law.

**Q 26. What is the principle involved in PDA method?**

- Nitrate reacts with PDA and produces a nitro derivative which in alkaline solution develops yellow colour due to rearrangement of its structure. The yellow colour follows Beer's law and is proportional to concentration of nitrate present in sample.

**Q 27. At which wavelength nitrate estimation is carried out in UV-Visible spectrophotometeric method?**

- Nitrate concentration is estimated at 220 nm and 275 nm in UV-Visible spectrophotometeric method.

**Q 28. At which wavelength nitrate concentration estimation is carried out in PDA method?**

- Nitrate concentration is estimated at 410 nm by spectrophotometer in PDA method.

**Q 29. Which regents are used in PDA method?**

- Reagents used in PDA method includes Phenol disulphonic acid, KOH (12 *N*) and stock nitrate solution (100 mg/L).

**Q 30. Give reaction involved in nitrification.**

- The reaction involved in nitrification is $NH_4 \longrightarrow NO_2 \longrightarrow NO_3$.

**Q 31. Which bacteria are responsible for nitrate fixation or nitrification?**

- Nitrifying bacteria are responsible for nitrate fixation or nitrification. Such as *Thiobacillus denitrificans*, *Micrococcus denitrificans* and some species of *Serratia*, *Pseudomonas* and *Achromobacter*.

**Q 32. Which bacteria are responsible for denitrification?**

- Denitrifying bacteria is responsible for denitrification.

**Q 33. What is the effluent discharge standard for nitrate?**

- According to MoEF (1993), 10 mg/L (inland surface water) and 20 mg/L (marine coastal areas) are effluent discharge standards for nitrate.

**Q 34. What is the stream standard for nitrate?**

- According to CPCB (1979), 20 mg/L (class A) and 50 mg/L (class C) are the standards for nitrate in stream.

**Q 35. How can the sample of nitrate be stored?**

- Sample of nitrate can be stored for longer duration by addition of 2 mL of conc. $H_2SO_4$ and stored at 4°C. When sample is preserved with acid, $NO_3$ and $NO_2^-$ cannot be determined individually.

**Q 36. How is a sample treated for nitrate estimation?**

- Sample is treated for nitrate estimation by filtering if necessary and adding 1 mL of conc. HCl and mixing thoroughly.

## 2.16 Dissolved Oxygen

**Q 1. What is dissolved oxygen (DO)?**

- The amount of oxygen in dissolved state in water is known as dissolved oxygen (DO).

**Q 2. What type of analysis is DO?**

- DO is a type of demand analysis.

**Q 3. Does sewage water contain dissolved oxygen?**

- Sewage water may contain very less quantity of dissolved oxygen.

**Q 4. Does surface water contain dissolved oxygen?**

- Yes, surface water contains dissolved oxygen.

**Q 5. What is the effect of DO on taste of water?**

- Water lacking DO has a dull taste and those saturated in DO has a pleasant taste.

**Q 6. For what purpose drinking water should be aerated?**

- Drinking water should be aerated to have a pleasant taste.

**Q 7. Water used for feeding boilers should have DO?**

- No, water used for feeding boilers should not have DO. It's presence in feeding water is undesirable and the risk of corrosion increases.

**Q 8. What is the effect of temperature on DO concentration?**

- The DO concentration decreases with increase in temperature.

**Q 9. Which apparatus is recommended for collection of DO sample?**

- BOD bottle is generally recommended for collection of DO sample.

**Q 10. What is the capacity of BOD bottle?**

- The capacity of BOD bottle is about 300 mL.

**Q 11. What is the maximum holding time for preserved DO sample?**

- Generally no storage is allowed but a time duration of 8 hours can be considered for sample analysis after preservation.

**Q 12. What is the range of DO with respect to aesthetic quality?**

- Till now, no guideline value has been set for DO with respect to aesthetic quality, but some standard indicates it as 4 mg/L.

**Q 13. What is the role DO in wastewater?**

- In wastewater DO is the factor that determines whether biological changes are brought out by aerobic or anaerobic microorganism.

**Q 14. How DO test is related to BOD test?**

- DO test is the basis of BOD test which is an important parameter to evaluate pollution potential of wastewater.

**Q 15. What are the methods for determination of DO?**

- There are two methods for determination of DO: Winkler method with Azide modification and Membrane electrode method.

**Q 16. What is the principle involved in Winkler method with Azide modification?**

- Oxygen present in water sample rapidly oxidizes the dispersed divalent magnous hydroxide to its higher valency which is precipitated as a brown hydrated oxide after addition of NaOH and KI. Upon acidification manganese reverts back to divalent state and liberates iodine from KI equivalent to original DO content. The liberated iodine is titrated against standard $Na_2S_2O_3 \cdot 5H_2O$ using starch as an indicator.

**Q 17. What is Winkler A solution?**

- Winkler A is a solution of manganese sulphate.

**Q 18. What is Winkler B solution?**

- Winkler B is a solution of alkali iodide azide.

**Q 19. What is alkali in Winkler B?**

- Alkali in Winkler B is sodium hydroxide (NaOH) or potassium hydroxide (KOH).

**Q 20. Which iodide salt is used in preparation of Winkler B?**

- Potassium iodide (KI) or sodium iodide (NaI) salt is used in reagent preparation of Winkler B.

**Q 21. What is sodium azide?**

- Sodium azide is $NaN_3$ (Sodium nitrite).

**Q 22. What is the role of concentrated $H_2SO_4$ in DO estimation?**

- Sulphuric acid reverts manganese from higher valency to its divalent state and liberates iodine from KI.

**Q 23. What is nitrogen fixation?**

- Settling time given to precipitate after addition of Winkler B is known as nitrogen fixation.

**Q 24. Which titrant is used in DO estimation?**

- The titrant used in DO estimation is sodium thiosulphate ($Na_2S_2O_3 \cdot 5H_2O$) which is also known as hypo solution.

**Q 25. What does end point indicates in DO estimation?**

- The end point in DO estimation indicates that liberated iodine was titrated against sodium thiosulphate ($Na_2S_2O_3 \cdot 5H_2O$).

**Q 26. What is the end point in Winkler method?**

- The end point in Winkler method is blue colour that changes to colourless.

**Q 27. By which name Winkler method was earlier known as?**

- Winkler method was earlier known as Alsterberg Azide modification method.

**Q 28. What are other methods for DO estimation?**

- Other methods for DO estimation includes:
  a) Rideal Stewart modification
  b) Alum flocculation modification
  c) Copper Sulphate-Sulfamic acid flocculation modification and
  d) Alkaline hypochlorite modification.

**Q 29. State chemical reaction after addition of Winkler A.**

- $MnSO_4 + KOH \longrightarrow Mn\,(OH)_2 + K_2SO_4$.

**Q 30. What does Mn $(OH)_2$ + $K_2SO_4$ indicate?**

- Magnous hydroxide + potassium sulphate indicates white precipitate.

**Q 31. What is the reaction after addition of Winkler B?**

- The reaction after addition of Winkler B is :

  $2Mn\,(OH)_2 + O_2 \longrightarrow 2MnO\,(OH)_2$.

  It indicates that divalent magnous hydroxide was excited to its higher state which was indicted by its brown precipitate.

**Q 32. What is the chemical reaction after addition of conc. $H_2SO_4$?**

- After addition of conc. $H_2SO_4$ manganese reverts to divalent state and liberates iodine from KI equivalent to the original DO content.

  $$MnO\,(OH)_2 + 2H_2SO_4 \longrightarrow Mn\,(SO_4)_2 + 3H_2O$$

  $$Mn\,(SO_4)_2 + 2KI \longrightarrow MnSO_4 + K_2SO_4 + I_2$$

**Q 33. What is the role of liberated iodine?**

- Liberated iodine gets titrated against standard $Na_2S_2O_3.5H_2O$ (N/80) using starch as an indicator.

**Q 34. Give the chemical reaction for the above step.**

- $Na_2S_2O_3.5H_2O + I_2 \longrightarrow Na_2S_4O_6 + 2NaI + 10H_2O$

**Q 35. What is the atomic number of oxygen?**

- The atomic number of oxygen is 8.

**Q 36. What is the atomic weight of oxygen?**

- The atomic weight of oxygen is 16.

**Q 37. What is the equivalent weight of oxygen?**

- The equivalent weight of oxygen is 8.

**Q 38. Give the formula for the calculation of dissolved oxygen.**

- The formula for the calculation of dissolved oxygen is
  DO = A × *N* × eq. wt. × 1000/Volume of sample in mL (200 mL).

**Q 39. What is A?**

- A is the volume of $Na_2S_2O_3.5H_2O$ required during titration.

**Q 40. What is *N*?**

- *N* is the normality of $Na_2S_2O_3{\cdot}5H_2O$ i.e. 0.025 *N*.

**Q 41. What volume of sample was taken as a titrant in DO estimation?**

- Volume of sample taken for titration in DO estimation was 201 mL which was correspond to 200 mL of original sample.

**Q 42. Why is the volume taken as 201 mL?**

- When 1 mL $MnSO_4$ followed by 1 mL alkali-iodine azide regent was added into the sample, 2 mL of original water sample was lost from BOD bottle. Therefore, 201 mL of sample was taken for titration which will correspond to 200 mL of original sample.

**Q 43. In which apparatus DO sample was collected?**

- DO sample was collected in BOD bottle.

**Q 44. What is the unit for reporting dissolved oxygen concentration?**

- Dissolved oxygen concentration is reported in mg/L or ppm.

**Q 45. Does DO concentration have relation with temperature?**

- Yes, DO concentration have relation with temperature. As temperature increases DO concentration decreases. DO is inversely proportional to temperature.

**Q 46. How can the drawback of Winkler method be overcome?**

- Drawback of Winkler method can be overcome by using membrane electrode method.

**Q 47. What is the advantage of membrane electrode method over Winkler method?**

- With membrane electrode system DO profile is useful for taking reading at streams or reservoirs itself.

**Q 48. What is a membrane electrode made up of?**

- Oxygen sensitive membrane electrodes are made of polarographic or galvanic type which composed of two solid metal electrodes in contact with supporting electrolyte separated from test solution by a sensitive membrane.

**Q 49. What is sensitive membrane made up of?**

- The sensitive membrane is made up of polyethylene and fluorocarbon membranes.

**Q 50. How are the membrane electrodes calibrated?**

- Usually membrane electrodes are calibrated by taking observations against air or sample of known DO concentration.

**Q 51. What is a diffusion current?**

- Diffusion current is linearly proportional to concentration of molecular oxygen.

## ■ 2.17 Chemical Oxygen Demand

**Q 1. What is Chemical Oxygen Demand (COD)?**

- Chemical oxygen demand in the quantity of a specified oxidant that reacts with a sample (water or wastewater) under controlled condition.

**Q 2. What is the unit of COD?**

- COD is expressed in mg/L or ppm.

**Q 3. Why is the measurement of COD carried out?**

- COD is often measured as a rapid indicator of organic pollution in wastewater. It is normally measured in both municipal and industrial wastewater treatment plant and gives an indication of the efficiency of the treatment process.

**Q 4. Name the method used for COD estimation.**

- COD can be measured by open / closed reflux titrimetric method and open / closed reflux colorimetric method.

**Q 5. Which reagents are used for determination of COD?**

- Reagents used for determination of COD includes:
  1) Potassium dichromate solution
  2) Sulphuric acid
  3) Standard Ferrous Ammonium Sulphate (FAS) (0.1 *N*)
  4) Ferrion indicator
  5) Mercuric sulphate and
  6) Silver sulphate.

**Q 6. What are interfering agents in COD estimation?**

- Fatty acids, straight chain aliphatic compounds, aromatic hydrocarbons, chlorides, nitrite and iron are interfering agents in COD estimation.

**Q 7. What is the normality of ferrous ammonium sulphate (FAS) used in COD estimation?**

- The normality of ferrous ammonium sulphate used in COD estimation 0.1 *N*.

**Q 8. What is the COD standard of drinking water?**

- There is no COD standard prescribed by WHO (1984), IS: 10500 (2012), USEPA (1974) or ICMR (1963).

**Q 9. What is the COD standard for effluent discharge?**

- The COD standard for effluent discharge is 250 mg/L into inland surface water and marine coastal area (MoEF1993; IS: 2490, 1981).

**Q 10. What are the applications of COD data?**

- 1) COD data gives an idea about the strength of the domestic and industrial wastewater.
  2) COD test is helpful in indicating toxic conditions and the presence of biologically resistance organic substances, and
  3) COD test is used widely in the operation of treatment facilities because of the speed with which results can be obtained.

**Q 11. Which method is used to determine COD?**

- COD test is carried out by Open Reflux method.

**Q 12. What does COD test determine?**

- COD test determines the oxygen required by wastewater for chemical oxidation of most organic matter and oxidizable inorganic substances with the help of strong chemical oxidant.

**Q 13. What are the advantages of COD over BOD?**

- COD determination has advantage over BOD determination as in COD results can be obtained in about 3 hours as compared to 5 days required for BOD. Further, COD test is relatively easy, gives reproducible results and is not affected by interferences as in BOD test.

**Q 14. What will be the effect of chlorides on COD results?**

- COD results will not be accurate if the sample contains more than 1000 mg/L chlorides.

**Q 15. In which sample reflux method is not applied?**

- Reflux method is not applied to samples such as sea water and brines.

**Q 16. Which titrant is used in determination of COD?**

- Ferrous ammonium sulphate of 0.1 *N* is used as a titrant in determination of COD.

**Q 17. What is the best use of COD test result?**

- One of the best uses of COD test result is rapid analysis of organically polluted samples.

**Q 18. What is the time required for COD sample digestion?**

- Time required for COD sample digestion is of about 2 hours.

**Q 19. Why $AgSO_4$ is added into concentrated $H_2SO_4$ during COD estimation?**

- Silver sulphate is added into concentrated $H_2SO_4$ during COD estimation as it acts as a catalyst and stimulates oxidation of straight chain aliphatic and aromatic compounds.

**Q 20. Which indicator is used in determination of COD?**

- Ferroin indicator is used in determination of COD.

**Q 21. What is the use of reflux condenser?**

- A reflux condenser is used to prevent loss of volatile organic compounds and further maintains the volume of water/wastewater sample.

**Q 22. Which oxidizing agent is used in determination of COD?**

- Potassium dichromate is used as an oxidizing agent in determination of COD.

**Q 23. What is the normality of ferrous ammonium sulphate used in determination of COD?**

- Normality of ferrous ammonium sulphate solution used in determination of COD is 0.1 *N*.

**Q 24. What is the formula to calculate COD?**

- The formula to calculate COD is:

  COD mg/L = (A-B) × *N* × 8 × 1000 / mL of sample taken where,

  A - mL of ferrous ammonium sulphate used for blank

  B - mL of ferrous ammonium sulphate used for sample

  *N* - Normality of FAS used

  8 - Milli equivalent weight of oxygen.

  The amount of sample taken in 150 mL (20 mL sample + 10 mL potassium dichromate solution + 30 mL conc. sulphuric acid and after reflux for 2 hours 90 mL distilled water was added through condenser).

**Q 25. During the addition of conc. $H_2SO_4$ into sample, Why it has to be stirred continuously?**

- During addition of conc. $H_2SO_4$ into sample, it has to be stirred continuously because slow addition of it along with stirring prevents fatty acid to escape out due to high temperature.

**Q 26. What is the colour change during the end point of the titration?**

- The end point of titration is blue green colour changing to wine red colour.

**Q 27. State the chemical reactions involved in estimation of COD.**

- 1) $2K_2Cr_2O_7 + 8H_2SO_4 \longrightarrow 2K_2SO_4 + 2Cr(SO_4)_3 + 8H_2O + 3O_2$

  2) $C_6H_{12}O_6 + 6O_2 \longrightarrow 6CO_2 + 6H_2O$

  $Cr_2O_7 + 6Cl^- + 14H^+ \longrightarrow 3Cl_2 + 2Cr^{+3} + 7H_2O$

  3) $Hg + 2Cl \longrightarrow HgCl_2$

## 2.18 Biochemical Oxygen Demand

**Q 1. What is Biochemical Oxygen Demand (BOD)?**

- Biochemical Oxygen Demand is the demand of oxygen required by organic matter present in sewage to degrade it into simple inorganic form.

**Q 2. For what purpose is BOD estimated?**

- The purpose of estimation of BOD is to measure oxygen consumed by bacteria during the process of oxidizing organic matter under aerobic conditions.

**Q 3. How many days are required to carry out BOD test? At which temperature test is carried out?**

- To carry out BOD test 5 days are required. The test is carried out at 20°C.

**Q 4. What is the efficiency for 5 days BOD test?**

- It has been proved by various experiments that 5 days BOD has about 68% efficiency.

**Q 5. What is the efficiency for 10 days BOD test?**

- Ten days BOD test has an efficiency of about 90%.

**Q 6. By how many methods BOD test can be carried out?**

- BOD test can be carried out by two methods: Direct method and Dilution method.

**Q 7. Which method is more efficient for estimation of BOD and why?**

- Dilution method is more efficient for estimation of BOD because it can easily carried out in laboratory and also is less expensive.

**Q 8. What is the BOD discharge standard in public sewers as per the Indian Standard Institute?**

- According to the Indian Standard Institute, BOD discharge should have a concentration of 350 mg/L in public sewers.

**Q 9. What is the discharge standard of BOD as per APHA on land for irrigation and for sewage surface?**

- According to APHA, the discharge standard for BOD is 100 mg/L for on-land for irrigation and 20 mg/L for sewage surface.

**Q 10. Why BOD concentration is always half or one-third of COD concentration?**

- BOD measures only bio-degradable organic matter hence its concentration is between half and one third of COD concentration.

**Q 11. Does BOD concentration decreased or increased after 5 days? If so why?**

- BOD concentration gets decreased after 5 days because dissolved oxygen in water or wastewater gets utilized by micro-organisms to decompose biodegradable organic matter into simple inorganic form.

**Q 12. Why BOD test is important in sewage analysis?**

- BOD test is important in sewage analysis as it grants uniformity while comparing various results and tests.

**Q 13. What is the concentration of BOD in strong sewage?**

- In strong sewage BOD has a concentration of about 450 to 550 mg/L.

**Q 14. What is the concentration of BOD in average sewage?**

- In average sewage BOD has a concentration of about 350 mg/L.

**Q 15. What is the concentration of BOD in weak sewage?**

- In weak sewage BOD has a concentration of about 250 mg/L.

**Q 16. What is the concentration of BOD in standard filter effluent?**

- In standard filter effluent BOD has a concentration of about 20 mg/L.

**Q 17. What is the concentration of BOD in a very good filter effluent?**

- In a very good filter effluent BOD has a concentration of about 5 to 10 mg/L.

**Q 18. In which country BOD test was discovered?**

- BOD test was discovered in the United Kingdom.

**Q 19. What happens to stoichiometric validity in BOD test?**

- The BOD test loses its stoichiometric validity after soluble organic matter present in the sample of sewage has been exhausted or utilized.

**Q 20. Which reagents are used in BOD estimation?**

- Reagents used in BOD estimation includes phosphate buffer, magnesium sulphate, calcium chloride and ferric chloride.

**Q 21. For 5 days BOD test, at which temperature sample is incubated?**

- For 5 days BOD test sample is incubated at 20°C in an incubator.

**Q 22. How is BOD calculated?**

- BOD is calculated by using the formula $BOD_{5\ days}$ mg/L = $D_1 - D_2 / \rho$ $\rho$ (Rho).

**Q 23. What is $D_1$ and $D_2$ in BOD test?**

- $D_1$ indicates initial dissolved oxygen i.e. on first day and $D_2$ indicates dissolved oxygen remained after incubation at 20°C on $5^{th}$ day.

**Q 24. What is Rho ($\rho$)?**

- Rho ($\rho$) is a volumetric fraction of water formula.

**Q 25. What is nutrient treated aerated distilled water?**

- Nutrient treated aerated distilled water is the water saturated with dissolved oxygen along with seed (microorganism) and all types of buffers.

**Q 26. What is seed water?**

- Sewage water containing different microorganism is known as seed water.

**Q 27. For what purpose buffers are added into BOD test?**

- The buffers are added into the BOD test to provide nutrients (food) for microorganism and to maintain pH of aerated water.

**Q 28. Is there any need to aerate seed water?**

- Yes, there is a need to aerate seed water, because sewage water hardly contains dissolved oxygen and for BOD test it is essential to aerate seed water.

**Q 29. For how long seed water has to be aerated?**

- Seed water has to be aerated for about 24-48 hours.

**Q 30. How much distilled water is required for preparation of dilution water?**

- Approximately three litres of distilled water is required for preperation of dilution water in BOD test.

**Q 31. What is dilution water?**

- The distilled water which is aerated (about 24-48 hours) along with seed water and buffers used for dilution purpose is known as dilution water. It is also known as nutrient treated aerated distilled water.

**Q 32. State the volume of seed water added in dilution water.**

- Generally 2 mL per litre of seed water is added in dilution water.

**Q 33. State the volume of buffers added in dilution water.**

- Buffer solutions are added in the quantity of one mL per litre i.e. 3 mL for 3 L dilution water.

**Q 34. Which type of sample is suitable for BOD test?**

- Sample having very high COD value is suitable for BOD test.

**Q 35. Why BOD sample is distributed as higher and lower concentration?**

- BOD sample is distributed as higher and lower concentration to ensure that demand does not increase the available oxygen.

**Q 36. What is the pH required during 5 days BOD test?**

- The pH required during 5 days BOD test is from 6.5 to 7.5.

**Q 37. What is the duration of aeration of dilution water?**

- The duration for aeration of dilution water is about 24 to 48 hours.

**Q 38. Why the dilution water is aerated?**

- Dilution water is aerated to satisfy oxygen demand of microbes.

**Q 39. Are there any obligations with dilution water?**

- Yes, there are obligations with dilution water. It may be tap or receiving stream water free of bio-degradable organism and bio-inhibitory substance such as chlorine or heavy metals.

**Q 40. What happens to DO concentration in BOD test?**

- Dissolved oxygen concentration in BOD test gets decreased after 5 days.

**Q 41. Is BOD calculation possible by any other formula?**

- Yes, the formula is BOD as $O_2$ mg/L = $(D_1 - D_2) - (B_1 - B_2) \times 100$ / percentage dilution.

**Q 42. What is $B_1$ and $B_2$ in the formula?**

- $B_1$ is DO of blank (seeded dilution water) before incubation and $B_2$ is DO of blank (seeded dilution water) after incubation.

**Q 43. What is seed control?**

- Determination of BOD of seeding material is known as seed control.

**Q 44. What should be the value of DO uptake of seeded dilution water?**

- The DO uptake of seeded dilution water should be between 0.6 mg/L and 1.0 mg/L.

**Q 45. From where the sewage sample for seed water should be collected?**

- Sewage sample for seed water can be obtained from the receiving water body preferably 3-8 km below the point of discharge.

**Q 46. Which instrument is used to maintain the temperature of 20°C?**

- Incubator is used to maintain the temperature of 20°C.

**Q 47. Give the general equation involved in BOD test.**

- The general equation involved in BOD test is:

$$\text{Sewage} + \text{bacteria} \longrightarrow \text{Treated waste} + \text{new bacteria.}$$

**Q 48. What is strong trade waste?**

- The effluents from industries having strong toxic chemicals are known as strong trade waste.

**Q 49. What is raw or settled sewage?**

- The sewage containing raw i.e. all insoluble impurities settling at the bottom is known as raw or settled sewage.

**Q 50. By which method DO is estimated in BOD test?**

- DO estimation in BOD test is carried out by Winkler method with Azide modification.

**Q 51. What is the significance of BOD test?**

- BOD test measures biodegradable organic matter and also essential in sewage treatment.

**Q 52. What is the disadvantage of BOD test?**

- BOD test has a lengthy procedure and time consuming.

## ■ 2.19 Oil and Grease

**Q 1. What is oil?**

- Oil is any material recovered as a substance soluble in solvent.

**Q 2. What is grease?**

- Grease is the material recovered as a substance soluble in solvent.

**Q 3. What happened when oil and grease are discharged into treated wastewater?**

- When oil and grease are discharged into treated wastewater they may cause surface films and shoreline deposits.

**Q 4. What are the methods used for analysis of oil and grease?**

- Generally three methods are used for analysis of oil and grease:
  a) The partition gravimetric
  b) The partition infrared method and
  c) The soxhlet extraction method.

**Q 5. Which method is used for analysis of oil and grease from liquid sample?**

- Partition gravimetric method is used for analysis of oil and grease from liquid sample.

**Q 6. Which method is designed for sample containing volatile hydrocarbon?**

- The partition infrared method is implied for analysis of sample containing volatile hydrocarbon.

**Q 7. Which method is suitable for determination of sludge?**

- The soxhlet method is suitable for determination of sludge.

**Q 8. Which reagents are required for partition gravimetric method?**

- The reagents required for partition gravimetric method includes:
  a) Conc. HCl
  b) Trichlorotrifluoroethane
  c) n-Hexane
  d) Methyl-tirtiary-butyl ether and
  e) Sodium sulphate crystals.

**Q 9. State the role of trichlorotrifluoroethane.**

- For the separation of sludge and oil from the sample trichloro-trifluoroethane is used.

**Q 10. What is 'known oil'?**

- A 'known oil' is defined as a sample of oil and or grease that represents only material of that type used or manufactured in the processes represented by a wastewater.

**Q 11. What is 'unknown oil'?**

- An 'unknown oil' is defined as one, for which a representative sample of oil or grease in not available for preparation of a standard.

**Q 12. By which formula oil and grease is estimated.**

- For calculation of oil and grease the formula is:

  Oil and grease mg/L = wt. of dish after drying - wt. of empty dish × 1000/ mL sample aliquot.

**Q 13. Which apparatus is used to weigh the sludge in wastewater sample?**

- Platinum dish is used for weighing sludge in wastewater sample.

**Q 14. How much quantity of trichlorotrifluoroethane is required for analysis?**

- For analysis, about 25-30 mL of trichlorotrifluoroethane is required.

**Q 15. How long does a sample in a separating funnel needs to be shaken?**

- Generally sample in a separating funnel needs to be shaken for 5-10 minutes.

**Q 16. Which instrument is used for drying the separated sample?**

- Preferably hot air oven is used for drying the separated sample.

**Q 17. What is the temperature required for drying a sample?**

- Temperature required for drying a sample is 105 to 110°C.

**Q 18. Why oil and grease are undesirable in sewage?**

- Oil and grease are undesirable in sewage because it tends to fill up the voids of filtering media and it forms objectionable deposits on sewers and treatment units.

**Q 19. What is soxhlet extraction?**

- Any oil and solid or viscous grease present are separated from liquid samples by filtration. This process takes place in Soxhlet apparatus so it is known as Soxhlet extraction.

**Q 20. What is the range for discharge of oil and grease into inland surface water according to IS: 2490-1982?**

- According to IS 2490-1982, range for discharge of oil and grease into inland surface water is 10 mg/L.

**Q 21. What is the range for discharge of oil and grease into land for irrigation?**

- The range for discharge of oil and grease into land for irrigation is 10 mg/L.

**Q 22. What is the maximum limit of oil and grease discharge into marine coastal area?**

- Maximum limit of oil and grease discharge into marine coastal area is 20 mg/L.

**Q 23. What is the standard for oil and grease discharge into sewers?**

- Discharge standard for oil and grease into sewers is 20 mg/L.

**Q 24. Name various parts of Soxhlet extraction assembly.**

- Soxhlet extraction assembly consist of porous thimble for placing solid, extraction tube, round bottom flask for solvent, condenser, side tube for solvent, siphon for condensed solvent, cotton plug and heating mantle.

**Q 25. What is the role of extraction tube?**

- Extraction tube is used for collection of oil and grease.

**Q 26. What is another formula for calculation of oil and grease?**

- Oil and grease (mg/L) = (A-B) × 1000/ mL of sample.

**Q 27. What is A?**

- It is the final weight of flask after solvent drying.

**Q 28. What is B?**

- It is the initial weight of empty flask.

**Q 29. Why heating mantel is used?**

- Heating mantel is used for heating purpose of organic layer.

**Q 30. Why sodium sulphate is used in oil and grease analysis?**

- Sodium sulphate is used in oil and grease analysis to remove moisture content present in it.

**Q 31. Which filter paper is used in oil and grease analysis?**

- Whatman No. 40 filter paper is used in oil and grease analysis.

**Q 32. What is the pore size of Whatman No. 40 filter paper?**

- The pore size of Whatman No. 40 filter paper is 0.06 mm.

**Q 33. Within two layers in separating funnel what the lower layer indicates.**

- Within two layers in separating funnel the lower layer indicates solvent layer.

**Q 34. What is the diameter of filter paper used for filtration?**

- The filter paper used for filtration purpose has a diameter of 110 mm.

**Q 35. How sample for oil and grease can be preserved?**

- Sample for oil and grease can be preserved by keeping it in refrigerator at 4°C.

**Q 36. What is the maximum holding time for oil and grease sample?**

- The maximum holding time for oil and grease sample is 6 hours.

**Q 37. Which type of container is recommended for oil and grease sample preservation?**

- A wide mouth glass type container is recommended for oil and grease sample preservation.

**Q 38. Which preservative is recommended for oil and grease sample preservation?**

- A concentrated sulphuric acid (36 *N*) is recommended for oil and grease sample preservation.

**Q 39. What pH should be maintained while preserving the sample?**

- While preserving the sample pH should be 2 or less than that.

**Q 40. What type of analysis is oil and grease?**

- Oil and grease is a type of organic analysis.

**Q 41. What is the role of acetone?**

- Acetone is added if the residue consists of visible water.

**Q 42. What volume of acetone was added into a sample?**

- Acetone at a rate of 2 mL per analysis was added into a sample.

## 2.20 Plankton

**Q 1. What is plankton?**

- Plankton is a free floating living organism.

**Q 2. Where plankton are found?**

- Plankton are found in all natural waters like lakes, rivers and in polluted water also.

**Q 3. In which different forms plankton are available?**

- Plankton are available broadly as phytoplankton and zooplankton.

**Q 4. What is phytoplankton?**

- If the component of plankton includes plant it is called as phytoplankton.

**Q 5. How does plankton move?**

- Plankton moves by using fine hair like projections around their body.

**Q 6. How are plankton collected?**

- Plankton are collected by using very fine net named as Planktonic net.

**Q 7. What is the pore size of Planktonic net?**

- The pore size of Planktonic net is 0.25 mm.

**Q 8. Which method is used for plankton collection?**

- Filtration method is used for plankton collection.

**Q 9. Are plankton producers?**

- Yes, phytoplankton are producers.

**Q 10. Does fish consume phytoplankton?**

- Yes, fish consume phytoplankton. Small fish consume phytoplankton because they are at the base of food chain.

**Q 11. Why plankton are important in ecosystem?**

- Plankton are important in food chain in pond ecosystem as they are the producer.

**Q 12. What is the difference between zooplankton and phytoplankton?**

- *Zooplankton :* They are microscopic creatures (animals) and utilize oxygen during respiration in water. They are heterotrophic in nature.

  *Phytoplankton :* They are microscopic plants and consume oxygen during respiration in water, also consume carbon dioxide during photosynthesis. They are autotrophic in nature. Phytoplankton are aquatic plant producer. They convert solar energy into chemical energy.

**Q 13. Are phytoplankton considered as autotrophs?**

- Yes, phytoplankton are autotrophic in nature as they can prepare their own food.

**Q 14. How does plankton float?**

- Phytoplankton are lighter than water as a result of which they can float on water.

**Q 15. Who consumes zooplankton?**

- Zooplankton are consumed by sea lions, manatees, sward fish etc.

**Q 16. What does absence of plankton indicates in a water body?**

- Absence of plankton in a water body indicates that water is polluted.

**Q 17. What does zooplankton consume?**

- Zooplankton consumes phytoplankton and decomposing matter and also microscopic aquatic animals.

**Q 18. How does zooplankton reproduce?**

- Zooplankton reproduces by asexual method.

**Q 19. Which apparatus is used for counting plankton?**

- Sedgwick Rafter Cell (S R Cell) is used as counting apparatus for phytoplankton and zooplankton both.

**Q 20. What is the dimension of Sedgwick Rafter Cell?**

- The dimension of Sedgwick Rafter Cell is 50 mm long, 20 mm width and 1 mm deep. The volume of the cell is 1 mL.

**Q 21. Why Sedgwick Rafter Cell is used for quantitative analysis of zooplankton?**

- Sedgwick Rafter Cell is commonly used for quantitative analysis of zooplankton because it can be easily operated and provide information when used with compound microscope.

**Q 22. What are the different categories of plankton?**

- Different categories of plankton includes: macro plankton, micro plankton and nano plankton.

**Q 23. What is the size of macro plankton?**

- The size of macro plankton is about 3 mm.

**Q 24. What is the size of micro plankton?**

- The size of micro plankton varies from 0.60 mm to 3 mm.

**Q 25. Give examples of nano plankton.**

- Examples of nano plankton includes bacteria and unicellular algae.

**Q 26. Which preservative is used for phytoplankton preservation?**

- Formaldehyde (4%) is used as a phytoplankton preservative.

**Q 27. Why and what volume of Lugol's iodine solution is added as a preservative?**

- Lugol's iodine solution is used as a preservative as it is least damaging for phytoflagellate. Lugol's iodine solution of 1 mL is to be added per 100 mL of water sample.

**Q 28. What is the function of preservative?**

- Lugol's iodine solution quickly does staining of phytoplankton and hence distinguishes them under microscope.

**Q 29. Which preservative should be used for zooplankton?**

- Zooplankton sample preservation is carried out immediately with 4% formaldehyde solution or Lugol's iodine solution.

**Q 30. What phytoplankton is comprised off?**

- Plankton that carries out photosynthesis are called phytoplankton and are comprised of organism called algae.

**Q 31. What is phytoplankton and what is their role in environment?**

- Phytoplankton is free floating flora, often minute organism that drift with water current. Phytoplankton produces oxygen and is important absorber of carbon dioxide and convert mineral to form that animals can utilize.

**Q 32. Why plankton are important?**

- As estimated 90% of all photosynthesis and production of usable oxygen take place in ocean. Plankton which are present in ocean thus plays a major role in oxygen production in ocean.

**Q 33. Give examples of phytoplankton.**

- Examples of phytoplankton include *Oscillatoria, Spirogyra, Ulothrix* and *Nostoc*.

**Q 34. Give examples of zooplankton.**

- Examples of zooplankton include *Daphnia, Cyclops, Cypris, Dioptomas* and *Keratella*.

**Q 35. Give examples of animal group in zooplankton.**

- Examples of animal group in zooplankton include *Protozoa, Rotifera* and *Copepoda*.

**Q 36. What are the characteristics of *Daphnia*?**

- The characteristics of *Daphnia* include:
    1) Body soft, laterally compressed measuring 1-2 mm.
    2) Segmentation is very important.
    3) Body except head is enclosed in a shell.
    4) A predominant median eye is present.

**Q 37. Which special microscope is used for quantitative estimation of plankton?**

- Inverted Planktonic Microscope is used for quantitative estimation of plankton.

3

# SOIL

## 3.1 Bulk and Particle Density

**Q 1. What is the bulk density of soil?**

- The bulk density of soil is defined as dry weight of a unit volume of it. This volume includes both solids and pores.

**Q 2. What is the abbreviation for bulk density of soil?**

- The abbreviation for bulk density is $D_b$.

**Q 3. What is the normal range of bulk density of soil?**

- Normally bulk density ranges from 1.1 to 1.5 g/cm$^3$ for medium to fine textured soil and 1.26 to 1.65 g/cm$^3$ for coarse textured soil. In alkaline soil slightly higher bulk density is observed.

**Q 4. What is the bulk density for sand and sandy loam soil?**

- Bulk density for sand and sandy loam soil ranges from 1.20 to 1.80 g/cm$^3$.

**Q 5. Why is bulk density measurement of soil important?**

- Bulk density measurement of soil is important as it shows the degree of compactness of soil structure and further used for calculating pore space of soil.

**Q 6. Establish relationship between bulk density and pore space of soil.**

- Bulk density is inversely related to pore space of soil.

**Q 7. How does high bulk density soil affects plant growth?**

- High bulk density soil affects plant growth by inhibiting root penetration which reduces access to groundwater and nutrients.

**Q 8. What is the general disadvantage of high bulk density soil?**

- General disadvantage of high bulk density soil is it has low permeability and infiltration.

**Q 9. State methods used for measurement of bulk density of soil.**

- Methods used for measurement of bulk density of soil include: Core sampler method (Bulk density of undisturbed soil) and Clod saturation method.

**Q 10. At what temperature wet core is dried in an oven?**

- Wet core is dried in an oven at 105°C for 24 hours.

**Q 11. In which unit soil bulk density is reported?**

- Soil bulk density is reported in g/cm$^3$.

**Q 12. What is the bulk density of very compact subsoil?**

- Very compact subsoil may have bulk density of 2.0 $g/cm^3$.

**Q 13. How does bulk density behaves with profile depth?**

- There is a distinct tendency for bulk density to rise with profile depth. This is due to the lower content of organic matter, less aggregation and root penetration and a compaction caused by weight of overlaying layers.

**Q 14. What will be the long term (40-100 years) effect of cropping on bulk density of topsoils?**

- Long term (40-100 years) effect of cropping will increase bulk density of topsoils.

**Q 15. What is particle density?**

- Particle density is a way of expressing soil weight in terms of density of solid particles making up soil. It is usually defined as the mass (or weight) of a unit volume of soil solids.

**Q 16. What is the abbreviation of particle density?**

- Particle density is abbreviated as $D_P$.

**Q 17. In which unit particle density is reported?**

- Particle density is reported in $mg/m^3$ (megagram per cubic meter).

**Q 18. If 1 $m^3$ of soil solids weights 2.4 mg what will be its particle density?**

- If 1 $m^3$ of soil solids weights 2.4 mg, then particle density will be 2.4 $mg/m^3$ (1 mg = 1 million grams and 1 $m^3$ = 1 million cubic centimetres).

**Q 19. What is the particle density of most mineral soil?**

- Particle density of most mineral soil usually varies between 2.60 and 2.75 $mg/m^3$.

**Q 20. Why mineral soil's particle density usually lies in the range of 2.60 to 2.75 $mg/m^3$?**

- Mineral soil's particle density usually lies in the range of 2.60 to 2.75 $mg/m^3$ because quartz, feldspar and colloidal silicates which usually make up the major portion of mineral soils had densities within this range.

**Q 21. Does size of particles of a given mineral and arrangement of soil solids have effect on particle density?**

- Size of particles of a given mineral and arrangement of soil solids have no effect on particle density.

**Q 22. Particle density depends upon which factor?**

- Particle density depends upon chemical composition and crystal structure of mineral particle.

## 3.2 Texture

**Q 1. What is soil texture?**

- Soil texture is the size distribution of soil particles.

**Q 2. Give names of textural classes of soil.**

- Textural classes of soil are sand, loamy sand, sandy loam, silt loam, sandy clay loam, silty clay loam, sandy clay, silty clay, silt and clay.

**Q 3. What is soil separates?**

- Sand, silt and clay are most commonly found in the solid state and are called soil separates.

**Q 4. What is soil aggregate?**

- The size of individual soil particles structure refers to the manner in which these individual soil particles are grouped together to form clusters of particles called soil aggregate.

**Q 5. What are the types of soil texture?**

- Soil texture is of three different types i.e. sand, silt and clay.

**Q 6. What is sand?**

- Sand particle consist of small pieces of primary unweathered rock fragments. They do not hold water.

**Q 7. What is the disadvantage of sand and gravel particles in soil?**

- The disadvantage of sand and gravel particles is that they do not supply any nutrient for crop growth as they do not store nutrients.

**Q 8. What is silt?**

- Silt particles are larger than clay particles but smaller than sand particles. They can hardly be seen by naked eye.

**Q 9. What is the advantage of silt soil texture?**

- Silt soil texture hold plant nutrients better than sand but not as well as clay. Silt particles allow water and air to pass through readily yet retain moisture for crop growth.

**Q 10. What is clay?**

- Clay particles are the smallest of all the particles found in soil. They cannot be seen by the naked eye. Many of them are too small to be seen under the microscope.

**Q 11. What is the advantage of clay particles?**

- Advantage of clay particles are that they are the most active part of soil and hold plant nutrients well. Clay has a high moisture retention capacity.

**Q 12. Name the methods used for determination of soil class.**

- The methods adopted for determination of soil class includes field method, mechanical analysis and international pipette method.

**Q 13. What is the principle involved in mechanical analysis of soil class?**

- The mechanical analysis of soil is based on the principle of fractionation of mineral components of the soil.

**Q 14. State the principle involved in international pipette method.**

- International pipette method is based on the principle of different sized particles having different weight fall at different velocities.

## ■ 3.3 Water Holding Capacity

**Q 1. What is called as water holding capacity of soil?**

- The specific ability of a particular type of soil to hold water against the force of gravity is called the water holding capacity of soil.

**Q 2. Which apparatus is used to determine water holding capacity of soil?**

- KR Box is used to determine water holding capacity of soil.

**Q 3. What is the full form of KR Box?**

- KR Box's full form is Keen Racz Kowski Box.

**Q 4. Which physical characteristics are observed in water holding capacity of soil?**

- Physical characteristics observed in water holding capacity of soil include:
  - i) Apparent density
  - ii) Absolute specific gravity
  - iii) Maximum water holding capacity and
  - iv) Percentage of pore space.

**Q 5. Water holding capacity of soil is reported in which unit?**

- Water holding capacity of soil is reported in percentage (%).

**Q 6. What is the formula to calculate water holding capacity of soil (WHC)?**

- $$\text{WHC }\% = \frac{(W_3 - W_1) \times (W_2 - W_1)}{(W_2 - W_1)} \times 100$$

**Q 7. What is meant by water logging?**

- The degree to which a soil becomes waterlogged depends on how much water enters into the soil and how quickly it leaves by deep percolation.

## 3.4 pH

**Q 1. What is pH?**

- pH is power of hydrogen ion activity OR pH is negative logarithmic value of hydrogen ($H^+$) ion activity pH = -log [$H^+$].

**Q 2. Which instrument is used to measure pH of soil?**

- pH meter is used to measure pH of soil.

**Q 3. pH meter electrodes comprises of.**

- pH meter electrodes comprises of glass electrode and calomel electrode.

**Q 4. Who discovered the pH scale?**

- Danish chemist P. L. Sorenson in 1909 discovered pH scale.

**Q 5. What is the formula to calculate pH of a solution?**

- The formula to calculate pH of a solution is :

$$pH = -\log_{10} [H^+] \text{ and } pOH = -\log_{10} [OH^-].$$

**Q 6. Which method is used to determine pH of soil?**

- Electrometric method is used to determine pH of soil.

**Q 7. Which fertilizer increases pH of soil or acidity of soil?**

- Ammonia based fertilizer has an acidifying effect on soil and are responsible for increase in pH of soil.

**Q 8. What is meant by soil pH?**

- Soil pH is referred as acidity of soil and is measured by number of hydrogen ions present in the soil solution.

**Q 9. What is the range of soil pH required for plant growth?**

- Soil pH range required for plant growth is about 6.5 to 7.2.

**Q 10. What is the composition of soil?**

- The composition of soil is: mineral matter 45%, organic matter 5%, water 25% and air 25%.

**Q 11. How is soil pH essential for plant nutrient?**

- Soil provides 14th of the 17th plant nutrients, hence soil pH is essential.

**Q 12. What is the range of pH scale?**

- The range of pH scale is from 0 to 14. Where from 0.0 to 6.9 is acidic, 7 is neutral and from 7.1 to 14.0 is alkaline.

**Q 13. Which force is produced in glass electrode at the time of determination of soil pH?**

- Electromotive force is produced in glass electrode at the time of determination of soil pH.

**Q 14. What is the significance of soil pH?**

- Most plant can tolerate wide pH range in solution but they cannot tolerate a wide range of acidity of soil. When soil acidity changes the solubility of the number of metal ions also change. Plant growth is affected by varying concentration of these metals in solution rather than by the acidity itself.

**Q 15. Which nutrients are available in soil and how is related to soil pH?**

- Nitrogen, P, K, Ca, Mg and S will have highest concentration in soil with pH 6-8. Concentration of available N is less sensitive to pH than concentration of available P. In order for P to be available for plants, soil pH needs to be in the range 6.0-7.5. If pH is lower than 6, P starts forming insoluble compounds with iron and aluminium. If pH is higher than 7.5, P starts forming insoluble compounds with Ca. Nutrient availability is in relation to soil pH.

**Q 16. What is the effect of water content on soil pH measurement?**

- In general, the more dilute the soil suspension, the higher the soil pH.

**Q 17. State advantages of glass electrode for measurement of soil pH.**

- Advantages of glass electrode for measurement of soil pH includes:
  1) It does not expel dissolved gases such as $CO_2$ from the system under measurement.
  2) It is not affected by oxidizing and reducing solution in contrast to quinohydrone or hydrogen electrode.
  3) It is adaptable to thick fluids, pastes, gels and coloured solutions.
  4) It is rapid, convenient, inexpensive and adaptable for continuous recording.

## 3.5 Conductivity

**Q 1. What is conductivity?**

- The ability or power to conduct electricity is called as conductivity.

**Q 2. Conductivity is measured by which instrument?**

- Conductivity is measured by conductivity meter.

**Q 3. What is the unit of conductivity?**

- The unit of conductivity is μSiemens/cm (μS/cm) or milliSiemens/cm (milliS/cm).

**Q 4. Who discovered the conductivity meter?**

- Alessandro Volta discovered the conductivity meter.

**Q 5. What are the contributors to hydraulic conductivity of soil?**

- The contributors to hydraulic conductivity of soil are gravel and coarse grained soil particles which are cohesion less.

**Q 6. What is meant by thermal conductivity of soil?**

- Thermal conductivity of soil is the ability to conduct heat.

**Q 7. What is the unit of thermal conductivity?**

- The unit of thermal conductivity is Watts per meter Kelvin (W/mK).

**Q 8. What is the unit of electrical conductivity?**

- The unit of electrical conductivity is milliSiemens per centimetre (mS/cm).

**Q 9. What is milliSiemens per centimetre?**

- MilliSiemens per centimetre is the standard unit of measurement of soil conductivity. A Siemens is a measurement of a materials conductance.

**Q 10. How is soil conductivity measured?**

- Soil conductivity can be measured by preparing a soil suspension by dissolving 1 part of soil to 1 part of distilled water to ensure no external ionic contamination. The conductivity electrode should be dipped into this soil suspension to measure conductivity. The soil conductivity result is reported along with weight/volume (w/v) suspension.

**Q 11. What is mean by high conductivity of soil?**

- High conductivity of soil indicates high concentration of ions or nutrients, therefore high soil salt content.

**Q 12. What is low conductivity of soil?**

- Low conductivity of soil is a sign of low ionic activity therefore low soil nutrient content i.e. the soil is empty.

## 3.6 Organic Carbon

**Q 1. Which methods are used for estimation of per cent organic carbon of soil?**

- Soil per cent organic carbon can be estimated by a) Dry combustion in a furnace or b) Chromic acid oxidation followed by measurement of $CO_2$ evolved.

**Q 2. What is the procedure for organic carbon determination in soil?**

- Organic carbon of soil is determined after removal or separate determination of carbonate, further subsequent ignition of soil in a furnace and determination of $CO_2$ evolved by a gravimetric or titrimetric method.

**Q 3. Which methods are available for determination of $CO_2$ evolved from soil organic carbon?**

- 1) Removal of carbonate and 2) Combustion method are used for determination of $CO_2$ evolved from soil organic carbon estimation.

**Q 4. State the importance of soil organic carbon determination.**

- The importance of soil organic carbon estimation is it provides an indication of organic matter content of soil which is an important index of soil fertility.

**Q 5. Which methods are used for soil organic carbon determination?**

- Dry combustion and wet digestion methods are used for determination of organic carbon in soil.

**Q 6. How is soil organic carbon oxidized to $CO_2$?**

- Soil organic carbon is oxidized to $CO_2$ in presence of an oxidizing agent such as potassium dichromate ($K_2Cr_2O_7$) and conc. $H_2SO_4$. The chromic acid formed by the action of $K_2Cr_2O_7$ and $H_2SO_4$ which oxidizes the carbon present in soil.

**Q 7. What is the end point of organic carbon determination?**

- Diphenyl amine indicator is used to detect colour change at the end point of organic carbon determination which is from blue to brilliant green.

**Q 8. Name the reagents used for determination of organic carbon of soil.**

- Reagents used for determination of organic carbon of soil include:
  1) Standard potassium dichromate solution
  2) Indicator (Any one) a) Diphenylamine indicator or b) Ferroin indicator
  3) Conc. $H_2SO_4$
  4) Orthophosphoric acid (85%) and
  5) Ferrous Ammonium Sulphate solution.

**Q 9. Name the method used for determination of per cent organic carbon in soil.**

- Walkley and Black method is used for determination of per cent organic carbon in soil.

**Q 10. In which unit, soil organic carbon is reported?**

- Soil organic carbon is reported in percentage (%) unit.

**Q 11. Which type of titration is employed in Walkley and Black method?**

- In Walkley and Black method back titration is employed.

**Q 12. Which is the oxidizing agent in Walkley and Black method?**

- Potassium dichromate ($K_2Cr_2O_7$) is the oxidizing agent in Walkley and Black method.

**Q 13. What happens when potassium dichromate and sulphuric acid is added into soil sample?**

- After addition of potassium dichromate and sulphuric acid into soil sample, the organic matter present in soil is broken down into $CO_2$ and $H_2O$. During this process dichromate is consumed.

**Q 14. What happen to the unutilized dichromate in the reaction?**

- The unutilized dichromate in the reaction is back titrated with ferrous ammonium sulphate solution.

**Q 15. What are the formulas for estimation of per cent organic matter?**

- There are two formulas for estimation of per cent organic matter:

  First formula is Organic carbon (%) = 10 (1-T/S) × 0.7792.

  Organic matter (%) = Organic carbon × 1.72.

  Where, 0.7792 is per cent recovery factor. This formula is used in case of ferrous ammonium sulphate is used as a titrant.

Second formula is Organic matter (%) = 10 (1-T/S) × 1.34. This formula is used in case of ferrous sulphate is used as a titrant.

**Q 16. How the factor 1.34 is derived?**

- The factor 1.34 is derived as

  (1.0 *N*) x 12/4000 × 1.72/0.77 × 100/0.5 = 1.34

  In which 0.5 is the sample weight (in gram), 1.72 the factor for organic matter from carbon and 12/4000 the meq weight of carbon. The 77 per cent is recovery factor founded by Walkley.

**Q 17. What is 1.72 in the formula?**

- 1.72 is the factor for organic matter from carbon is based on the assumption that carbon is only 58% of the organic matter.

## ■ 3.7 Chlorides

**Q 1. What is chloride?**

- The chloride ion is formed when the element chlorine, a halogen, gains an electron to form an anion (negatively charged).

**Q 2. What is the function of chloride in plant?**

- Functions of chloride in plant include: i) Chloride is essential for many plant functions, ii) Chloride also functions in photosynthesis, specifically in water splitting system, iii) Chloride is important in opening and closing of stomata and iv) Chemically balance the potassium ion ($K^+$) concentration.

**Q 3. Which factors affects chloride availability in soil?**

- Most soil chloride is highly soluble and is found predominantly dissolved in soil water.

**Q 4. Which method is used to determine chloride of soil?**

- Mohor's method is used to determine chloride content of soil.

  *(For further questions and answers of this section please refer section of chlorides in water).*

  *(For other soil related parameters please refer water section and parameters in it for basic questions and answers).*

4

# NOISE

**Q 1. What is acoustic?**

- Acoustic is the science of sound.

**Q 2. What is noise?**

- Noise is unwanted sound at wrong place and wrong time.

**Q 3. What is white noise?**

- A white noise is a sound similar to radio static or a sound made by a fan. This is often used to mask unpleasant sounds.

**Q 4. What is pink noise?**

- Pink noise or 1/f noise in it each octave carries an equal amount of noise energy. The name arises from the pink appearance of visible light with this power spectrum. It is random noise having equal energy per octave and so having more low frequency components than white noise.

**Q 5. What are the sources of noise?**

- Sources of noise can be broadly classified into: industrial and non-industrial.

**Q 6. Which instruments are used for measurement of noise?**

- Instruments used for measurement of noise includes sound level meter, pen recorder, cassette recorder, etc.

**Q 7. What is octave band analyzer?**

- Octave band analyzer is special sound level meter that divide noise into frequency components. Electronic filter circuits are use to divide the sound or noise into individual frequency bands.

**Q 8. How sound level meter is calibrated?**

- Sound level meter is calibrated by using acoustic calibrator. The acoustic calibrator fits over the microphone and outputs a consistent sound level for the microphone to detect.

**Q 9. In which unit noise is reported?**

- Noise is reported in decibel (dB) unit named after Sir Alfred Bell.

**Q 10. What does decibel refers to?**

- Decibel refers to the sound pressure level (SPL).

**Q 11. In which unit loudness is also expressed?**

- Loudness is also expressed in Sons.

**Q 12. What is one Son?**

- One Son is equal to the loudness of 40 dB sound pressure of 1000 Hz.

**Q 13. What is Phon?**

- Sound expressed in physcho-acoustic terms is called as Phon.

**Q 14. What is the loudest sound possible?**

- Normal air at sea level is 14.7 pounds per square inch (psi) or 1,01,325 Pascal (Pa), which is equivalent to 194 dB (A). So, 194 dB (A) is the loudest sound possible in air at sea level, beyond which it becomes a shock wave.

**Q 15. What is the threshold of hearing?**

- Threshold of hearing is zero dB (A).

**Q 16. What is threshold of pain?**

- Threshold of pain is 180 dB (A).

**Q 17. In which weighting unit noise is reported?**

- Noise is reported in 'A' weighting.

**Q 18. At which position from ground level noise monitoring is carried out?**

- Noise monitoring from ground level is carried out at receiver's position i.e. about 1.2 meter (at human ear's position).

**Q 19. For constant noise level monitoring which response time may be used?**

- For constant noise level monitoring a fast response time may be used.

**Q 20. For short time noise monitoring which response time may be used?**

- For short time noise monitoring (few seconds or minute) a slow response time may be used.

**Q 21. For measurement of noise on a highway at what distance noise level monitoring should be carried out?**

- For measurement of noise on a highway noise level monitoring should be carried out at a distance of 7.5 meters from the centre of a highway.

**Q 22. What is Leq?**

- Leq stands for equivalent sound pressure level which is sound pressure level averaged over the measurement period.

**Q 23. What is Ln?**

- Ln stands for noise level during night time. The night time is considered as from 10 pm to 6 am (nine hours).

**Q 24. What is Ld?**

- Ld stands for noise level during day time. The day time is considered as from 6 am to 10 pm (fifteen hours).

**Q 25. What is Ldn?**

- Ldn stands for noise level during day and night time (24 hours).

**Q 26. At what frequency noise level should be monitored?**

- Noise level should be monitored at a frequency of 24 hours of a day and at each hour frequency.

**Q 27. According to NAAQS for noise what are the different categories/zones of an area?**

- According to NAAQS for noise, four categories/zones of an area are identified. These include industrial, residential, commercial and silence zones.

**Q 28. What is a silence zone?**

- Silence zone is an area comprising not less than 100 meters around hospitals, educational institutions and courts. The silence zones are the zones which are declared by the competent local authority.

**Q 29. What is day time according to NAAQS w.r.t. noise reckoned as?**

- According to NAAQS w.r.t. noise day time is reckoned as time between 6 am to 22 pm (6 am to 10 pm), total 15 hours.

**Q 30. What is night time according to NAAQS w.r.t. noise reckoned as?**

- According to NAAQS w.r.t. noise night time is reckoned as time between 22 pm to 6 am (10 pm to 6 am), total 9 hours.

**Q 31. What is NAAQS w.r.t. noise for industrial zone?**

- According to NAAQS w.r.t. noise for industrial zone for day time is 75 dB (A) and for night time is 70 dB (A).

**Q 32. What is NAAQS w.r.t. noise for commercial zone?**

- According tc NAAQS w.r.t. noise for commercial zone for day time is 65 dB (A) and for night time is 55 dB (A).

**Q 33. What is NAAQS w.r.t. noise for residential zone?**

- According to NAAQS w.r.t. noise for residential zone for day time is 55 dB (A) and for night time is 45 dB (A).

**Q 34. What is NAAQS w.r.t. noise for silence zone?**

- According to NAAQS w.r.t. noise for silence zone for day time is 50 dB (A) and for night time is 40 dB (A).

**Q 35. What are occupational noise standards?**

- Occupational noise standards are those standards that are associated with occupation related noise.

**Q 36. Are occupation related noise standards and NAAQS w.r.t. noise are same.**

- Occupation related noise standards and NAAQS w.r.t. noise are not same, they are different.

**Q 37. Which agency has set up occupation related noise standards?**

- Occupation related noise standards were set up by OHSA (Occupational Health Safety Administration).

**Q 38. Which agency has set up NAAQS w.r.t. noise and when?**

- NAAQS w.r.t noise was setup by Central Pollution Control Board, New Delhi in the year 1986.

**Q 39. Is noise is affected by meteorological parameters?**

- Yes, noise is affected by meteorological parameters.

**Q 40. Name the meteorological parameters which affect noise?**

- Meteorological parameters which affect noise include wind speed and direction and humidity.

**Q 41. What TNI stands for?**

- TNI stands for Traffic Noise Index.

**Q 42. State the formula for calculating TNI.**

- For calculating Traffic Noise Index the formula is :

$$TNI = L_{90} + 4\ (L_{10} - L_{90}) - 30.$$

**Q 43. State the formula for calculating Ldn.**

- $Ldn = 10 \log_{10} \{1/24\ [15\ (10^{Ld/10}) + 9\ (10^{Ln+10/10})]\}.$

**Q 44. Which software is used for noise analysis?**

- There are number of software's developed by different agencies for noise analysis, one of them is *Dhwani* developed by NEERI, Nagpur.

**Q 45. What are the methods for noise pollution control?**

- Noise pollution can be controlled by technology and management aspects individually or by combination of these together with additional measures such as legislation, imposition of penalty, etc.

**Q 46. State technological interventions used for noise control.**

- Technological interventions used for noise pollution control includes noise control at source, along sound path way and at receiver end.

**Q 47. What are noise barriers?**

- Noise barriers are solid obstructions which can be constructed between highway and houses along the highway.

**Q 48. Which materials can be used in noise barriers?**

- Noise barriers can be constructed from earth, concrete, masonry, wood, glass, metal or other suitable material.

**Q 49. To what extent noise is attenuated by vegetation.**

- Noise can be attenuated by vegetation to an extent of 10 dB (A) per 100 meters width.

**Q 50. Name plant species which can be used in vegetative noise barriers.**

- Plant species which can be used in vegetative noise barriers include: Small size trees: Malbar nut (*Adhatoda vasica*), Sugar apple (*Annona squamosa*), Big sage (*Lantana camara*), Thuja (*Thuja orientalis*); Medium size trees: Indian gum arabic (*Acacia nilotica*), Flame of the forest (*Butea *monosperma*), Golden rain tree (*Cassia fistula*), Chinaberry tree (*Melia azedarach*); Large size trees: Neem (*Azadiracta indica*), Banyan (*Ficus benghalensis*), Teak (*Tectona grandis*), Bahera (*Terminalia bellerica*) and Arjuna (*Terminalia arjuna*); Creeper: Bougainvillea (*Bougainvillea spectabilis*).

**Q 51. How cumulative impact of multiple stationary noise sources (through hemispherical wave propagation) at a particular place can be calculated?**

- The cumulative impact of multiple stationary noise sources (through hemispherical wave propagation) at a particular place can be calculated by the formula:

$$L_P(\text{TOTAL}) = \underset{i=1}{\overset{N}{10}} \text{ Log}\left[\sum 10^{(L_{pi}/10)}\right]$$

where, $L_P$ (TOTAL) is the resultant cumulative sound pressure level due to N separate sources.

**Q 52. How impact from a continuous stationary noise source at observer's place, considering the hemispherical propagation of sound waves, can be calculated?**

- The impact from a continuous stationary noise source at observer's place, considering the hemispherical propagation of sound waves can be calculated by the following equation:

$$L_R = L_S + DI - 20\text{ Log}(r) - A_e - 8$$

Where,

$L_R$ : Sound pressure level at a receptor located at radial distance 'r', dB(A)

$L_S$ : Sound pressure level at the source, dB(A)

DI : Directivity index of the source [for hemispherical radiation DI= 3 dB(A)]

r : Radial distance of the receptor from the source, m

'$A_e$' : represents excess attenuation of sound caused by the environmental conditions such as:

- Absorption in Air
- Effect of rain, sleet, snow and fog
- Reflections and refractions at barriers and buildings
- Effect of the terrain, grass, shrubs, trees, etc.
- Effect of wind and temperature gradient

The absorption of sound is generally a complex phenomenon. In general, for homogeneous loss free atmosphere $A_e = 0$.

**Q 53. How is the equivalent noise level due to road traffic estimated using Federal Highway Administration (FHWA) noise model?**

- The equivalent noise level due to road traffic is estimated using the following Federal Highway Administration (FHWA) noise model:

$L_{eq}(h)_i = L_{Oei} + 10 \log [N_i/(S_i \times T_i)] + 10 \log (15/d)^{(1+a)} + S_o - 13$

$L_{eq}(h)_i$ = $L_{eq}$ at hour h for $i^{th}$ vehicle type, i.e. autos (LMV), medium trucks (MMV) or heavy trucks (HMV)

$L_{Oei}$ = The reference mean energy level for the $i^{th}$ vehicle type

$N_i$ = The number of $i^{th}$ class vehicles passing during the time T

$S_i$ = The average speed of the $i^{th}$ class vehicles, km/hr

T = The duration for which $L_{eq}$ is desired and must correspond to $N_i$ the count of vehicles during the time T. Normally, T would be one hour but can be longer or shorter

D = The perpendicular distance, in meters, from the centerline of the traffic lane to the location of the receptor, i.e. the location where the noise level is desired

a = The factor which relates to the absorption characteristics of the ground cover between the roadway and the receptor

$S_o$ = The shielding factor provided by a noise barrier, if present.

**Q 54. How cumulative noise effect from light, medium and heavy vehicles on the roads can be calculated?**

- Cumulative noise effect from light, medium and heavy vehicles on the roads can be calculated by using the following equation:

$$Leq\ (Total) = 10 \log [10^{Leql/10} + 10^{Leqm/10} + 10^{Leqh/10}]$$

Where, Leql, Leqm and Leqh are equivalent noise levels for light, medium and heavy vehicles respectively.

# *Appendix I*

**Table 1.** International Standard (IS) units

| | Unit | Symbol | in kg |
|---|---|---|---|
| **Mass** | Metric Tonne | MT | $1 \times 10^3$ |
| | Kilogram | kg | 1 |
| | Gram | g | $1 \times 10^{-3}$ |
| | Milligram | mg | $1 \times 10^{-6}$ |
| | Microgram | µg | $1 \times 10^{-9}$ |

| | Unit | Symbol | in meter |
|---|---|---|---|
| **Length** | Kilometre | km | $1 \times 10^3$ |
| | Meter | m | 1 |
| | Centimetre | cm | $1 \times 10^{-2}$ |
| | Millimetre | mm | $1 \times 10^{-3}$ |
| | Micrometer | µm | $1 \times 10^{-6}$ |
| | Nanometre | nm | $1 \times 10^{-9}$ |

| | Unit | Symbol | in litre |
|---|---|---|---|
| **Volume** | Kiloliter | kL | $1 \times 10^3$ |
| | Litre | L | 1 |
| | Millilitre | mL | $1 \times 10^{-3}$ |
| | Microlitre | µL | $1 \times 10^{-6}$ |

| | Unit | Symbol | in $mm^2$ |
|---|---|---|---|
| **Surface area** | Square kilometre | $km^2$ | $1 \times 10^6$ |
| | Hectare | ha | $1 \times 10^4$ |
| | Square meter | $m^2$ | 1 |
| | Square centimetre | $cm^2$ | $1 \times 10^{-4}$ |
| | Square millimetre | $mm^2$ | $1 \times 10^{-6}$ |

| | Unit | Symbol | in seconds |
|---|---|---|---|
| **Time** | Day | d | 86400 |
| | Hour | h | 3600 |
| | Minute | m | 60 |
| | Second | s | 1 |

# Appendix II

Table 2. Strength of common acids and bases

| Acid and Base | Formula | Specific gravity (Approx.) | Normality | % by weight |
|---|---|---|---|---|
| Sulphuric acid | $H_2SO_4$ | 1.84 | 36 | 98 |
| Nitric acid | $HNO_3$ | 1.42 | 16 | 70 |
| Hydrochloric acid | HCl | 1.19 | 11.6 | 37 |
| Liquor ammonia (Ammonium hydroxide) | $NH_4OH$ | 0.90 | 15 | 28 |
| Phosphoric acid | $H_3PO_4$ | 1.71 | 45 | 85 |

Table 3. Equivalent weight of substances in environmental analysis

| Name | Formula | Equivalent weight |
|---|---|---|
| Ammonium hydroxide (Liquor ammonia) | $NH_4OH$ | 35 |
| Ferrous ammonium sulphate (FAS, Mohr's salt) | $FeSO_4 (NH_4)_2 SO_4 \cdot 6H_2O$ | 392 |
| Hydrochloric acid | HCl | 36.5 |
| Nitric acid | $HNO_3$ | 63 |
| Potassium dichromate | $K_2Cr_2O_7$ | 49 |
| Potassium hydroxide | KOH | 56 |
| Silver nitrate | $AgNO_3$ | 170 |
| Sodium bicarbonate | $NaHCO_3$ | 84 |
| Sodium carbonate | $Na_2CO_3$ | 53 |
| Sodium chloride | NaCl | 58.5 |
| Sodium hydroxide | NaOH | 40 |
| Sodium thiosulphate (Hypo) (Anhydrous) | $Na_2S_2O_3$ | 158 |
| Sodium thiosulphate (Hypo) (Hydrated) | $Na_2S_2O_3 \cdot 5H_2O$ | 248 |
| Sulphuric acid | $H_2SO_4$ | 49 |

# Appendix III

**Table 4.** Methods of measurment of parameters and pollutants

| Parameter | Measurement technique |
|---|---|
| Meteorological parameters | |
| Wind speed, wind direction, ambient temperature | Mechanical weather monitoring station |
| Air pollutant | |
| Suspended particulate Matter (Particulate size 0.5 μ to 100 μ) | High volume air sampler (Gravimetric) |
| Respirable Particulate Matter (< 10 μ) | High volume air sampler with cyclone (Gravimetric) |
| Sulphur Dioxide | Modified West and Gaeke Method (Spectrophotometric) |
| Oxides of nitrogen ($NO_x$) in the form of Nitrogen dioxide ($NO_2$) | Sodium arsenite modified Jacob and Hochheiser Method (Spectrophotometric) |
| Carbon monoxide | Non Dispersive Infra Red Technique |
| Hydrogen sulphide | Methylene blue method (Spectrophotometric) |
| Ammonia | Nessler's Method (Spectrophotometric) |
| Hydrocarbon | Flame ionization detection and gas chromatography |
| Noise | Sound level meter |
| Water | Standard Methods for Examination of Water and Wastewater |
| Soil | Reference book by M. L. Jackson<br>Reference book by C. A. Black |
| Biological environment | Standard Methods for Examination of Water and Wastewater<br>Secondary data from Govt. offices and published literature<br>Flora: sample plot method, Transect and line intercepts method, Plotless sampling method<br>Fauna: Animals species list, count indices |
| Socio-economic environment | Secondary data from census records, statistical handbook, toposheets, health records and other relevant official records available with government agencies<br>Socio-economic survey |
| Topography and land use | Toposheets and remote sensing analysis techniques |
| GIS & Remote Sensing | Remote sensing analysis techniques |

# Appendix IV

**Table 5.** Classification of meteorological parameters

| Primary meteorological parameters | Secondary meteorological parameters |
|---|---|
| Wind | Atmospheric pressure |
| Turbulence and Atmospheric stability | Precipitation, Solar radiation |
| Mixing height, Ventilation coefficient | Temperature, Relative humidity |

**Table 5.** Insolation category

| Insolation category | Insolation (Langley/hour) |
|---|---|
| Strong | $R > 50$ |
| Moderate | $50 > R > 25$ |
| Weak | $25 > R > 12.5$ |
| Night | $R < 12.5$ |

R is the insolation. 1 Langley = 1 calorie per square centimetre

**Conversion of wind data from m/s to km/hr:** Whenever wind speed data is present in m/s and has to be converted into km/hr, multiply the value of m/s by 3.6 so as to get the data into km/hr.

# Appendix V

**Table 7.** Methods used for air environment study

| Parameter | Methods of measurement | Sampling flow rate | Sensitivity |
|---|---|---|---|
| Suspended Particulate Matter (SPM) | High volume air sampler<br>Glass micro fibre filter of 0.203 m × 0.254 m<br>Gravimetric method | 1.0-1.5 $m^3$/min | <10 µg/$m^3$ |
| Respirable Particulate Matter (RPM) | High volume air sampler with cyclone separator<br>Glass micro fibre filter of 0.203 m × 0.254 m | 1.0-1.5 $m^3$/min | <10 µg/$m^3$ |
| Sulphur Dioxide ($SO_2$) | High volume air sampler and gaseous sampling kit<br>Sampling through absorbing media in impinger<br>Wet chemical method (Modified West and Gaeke Method) | 0.5-1.0 L/min | 2 µg/$m^3$ |
| Oxides of Nitrogen ($NO_X$) in the form of Nitrogen dioxide ($NO_2$) | High volume air sampler and gaseous sampling kit<br>Sampling through absorbing media in impinger<br>Wet chemical method (Modified Jacob and Hochheiser Method) | 0.5-1.0 L/min | 3 µg/$m^3$ |
| Carbon monoxide (CO) | Tedlar bags<br>ND/IR/Gas filter correlation | Grab air sample (5 L) | 0.01 ppm |
| Heavy metals in Suspended Particulate Matter (SPM) | High volume air sampler<br>Glass micro fibre filter of 0.203 m × 0.254 m (EPM 2000)<br>Atomic Absorption Spectrophotometer (AAS) | 1.0-1.5 $m^3$/min | 0.01µg/$m^3$ |
| Stack monitoring Particulate matter | Stack sampling kit<br>Cellulose thimble of 30 mm × 100 mm | Isokinetic sampling | < 1 mg/N$m^3$ |

| Parameter | Methods of measurement | Sampling flow rate | Sensitivity |
|---|---|---|---|
| Sulphur Dioxide ($SO_2$) | Stack sampling kit with gaseous sampling kit<br>Sampling through absorbing media in impinger<br>Wet chemical method (Barium perchlorate) | 2 L/min | < 1 mg/$Nm^3$ |
| Oxides of nitrogen in the form of Nitrogen Dioxide ($NO_2$) | Stack sampling kit with gaseous sampling kit<br>Sampling through absorbing media in impinger<br>Wet chemical method (Phenol disulphonic acid) | 2 L/min | < 1 mg/$Nm^3$ |

# Appendix VI

**Table 8.** Air quality monitoring methods

| No. | Parameter | Method |
|---|---|---|
| 1 | Suspended Particulate Matter (μg/m$^3$) | Gravimetric method. Minimum detectable limit <10 μg/m$^3$ |
| 2 | Respirable Particulate Matter (μg/m$^3$) | Gravimetric method. Minimum detectable limit <10 μg/m$^3$ |
| 3 | Sulphur dioxide ($SO_2$) (μg/m$^3$) | Modified West and Gaeke method (KTCM method) |
| 4 | Oxides of nitrogen ($NO_2$) (μg/m$^3$) | Modified Jacob and Hochheiser method |
| 5 | Ammonia (μg/m$^3$) | Nesslerization method |
| 6 | Aldehyde (μg/m$^3$) | MBTH (3 methyl-2 benzothiazolone hydrazone hydrochloride) |
| 7 | Hydrocarbon (ppm) | Hydrocarbon analyzer (FID)/ GCMS |
| 8 | Hydrogen sulphide ($H_2S$) (μg/m$^3$) | Methylene blue |
| 9 | Sulphur dioxide from stack | BPA method (Barium perchlorate) |
| 10 | Oxides of nitrogen from stack | PDA method (Phenol disulphonic acid) |
| 11 | Benzo [a] pyrene (PAH) | Gas chromatography |

**Table 9.** Methods for air pollutants analysis

| No. | Pollutant | Methods of analysis |
|---|---|---|
| 1 | Suspended Particulate Matter | • High Volume Air Sampler (HVS)<br>• Nephelometric method<br>• Wind direction method<br>• Cascade impactor method |
| 2 | Sulphur dioxide ($SO_2$) | • Colorimetric method or Tetrachloromercurate<br>• Manual conductivity method<br>• Automatic conductivity method<br>• $SO_2$ monitoring with amperometric instruments<br>• Pulse fluorescence technique (0-100 ppm range; 0.001 ppm minimum delectable limit) |
| 3 | Oxides of nitrogen ($NO_x$) | • Chemiluminescence technique (0-100 ppm range; 0.001 ppm minimum detectable limit)<br>• Colorimetric method (Sodium Arsenite Method) |
| 4 | Carbon monoxide | • Manual colorimetric method<br>• Infrared absorption method<br>• Non-Dispersive infrared method<br>• Hopcolite method |

| No. | Pollutant | Methods of analysis |
|---|---|---|
| 5 | Hydrocarbon | • Gas chromatography<br>• Soxhlet extraction<br>• Column chromatography<br>• UV Visible spectrophotometer<br>• Fluorescence analysis<br>• FID (methane only, when charcoal column is present)<br>• ECD (ambient air sample analysis) |
| 6 | Ammonia | • Spectrophotometric<br>• Potentiometric (0.001 ppm) |
| 7 | Volatile pesticides | • Gas chromatography |

# Appendix VII

**Table 10.** WHO Standards for heavy metals in suspended particulate matter (SPM) (1987)

| Metal | Concentration |
|---|---|
| Cadmium | 1-5 ng/m$^3$ – rural for 1 year<br>10-20 ng/m$^3$ – urban for 1 year |
| Lead | 0.5-1.0 µg/m$^3$ for 1 year |
| Manganese | 1.0 µg/m$^3$ for 1 year |
| Nickel | Carcinogen |

**Table 11.** Settling velocity of unit density spheres in still air

| Particle diameter, µ | Settling velocity, Vs, cm/sec |
|---|---|
| 1000 | 385 |
| 500 | 200 |
| 220 | 76 |
| 100 | 25.1 |
| 50 | 7.2 |
| 40 | 4.8 |
| 30 | 2.7 |
| 10 | 0.30 |
| 1 | 0.0035 |

# Appendix VIII

**Table 12.** Classification of water analysis methods

| No. | Types of analysis | Parameters involved |
|---|---|---|
| 1 | Physical analysis | pH, turbidity, total suspended solids, total dissolved solids, conductivity and temperature. |
| 2 | Inorganic analysis | Alkalinity, acidity, total hardness, calcium hardness, chlorides, sulphate, sodium and potassium. |
| 3 | Nutrient analysis | Phosphate, nitrate, oil and grease, sulphide, phenol, hydrocarbon and total nitrogen. |
| 4 | Demand analysis | Dissolved oxygen, chemical oxygen demand and biochemical oxygen demand. |
| 5 | Heavy metals analysis | Fe, Ni, Co, Pb, Cu, Cr, Zn, Mn, Cd, etc. |

**Table 13.** Water quality analysis methods

| No. | Parameter | Method |
|---|---|---|
| 1. | pH | Electrometric method |
| 2. | Temperature | Mercury thermometer |
| 3. | Turbidity (NTU) | Nephelometeric/Turbidimetric method |
| 4. | Total dissolved solids (mg/L) | Gravimetric method |
| 5. | Total suspended solids (mg/L) | Gravimetric method |
| 6. | Conductivity (μS/cm) | Conductivity meter |
| 7. | Total alkalinity (mg/L as $CaCO_3$) | Acid base titration method |
| 8. | Acidity (mg/L as $CaCO_3$) | Acid base titration method |
| 9. | Total hardness (mg/L as $CaCO_3$) | EDTA complexometric titration |
| 10. | Chlorides (mg/L) | Argentometric titration (Mohor's method) |
| 11. | Sulphide (mg/L) | Iodometric titration |
| 12. | Sulphate (mg/L) | Turbidimetric method |
| 13. | Sodium (mg/L) and potassium (mg/L) | Flame photometer/ICP/AAS |
| 14. | Nitrate (as N, mg/L) | UV Visible spectrophotometric screening method |
| 15. | Total phosphate (mg/L) | Stannous chloride method |
| 16. | Dissolved oxygen (mg/L) | Winkler method with azide modification |
| 17. | Chemical oxygen method (mg/L) | Open reflux method |
| 18. | Biochemical oxygen demand (mg/L) | 5 days BOD at 20°C |

| No. | Parameter | Method |
|---|---|---|
| 19. | Nickel | Acid digestion method followed by ICP/AAS |
| 20. | Cadmium | |
| 21. | Chromium | |
| 22. | Cobalt | |
| 23. | Lead | |
| 24. | Iron | |
| 25. | Manganese | |
| 26. | Zinc | |
| 27. | Nitrogen (ammonia) (mg/L) | Kjeldahl method |
| 28. | Nitrogen (nitrate) (mg/L) | Colorimetric, Brucine method |
| 29. | Residual free chlorine (mg/L) | Iodometric titration |
| 30. | Oil and grease (mg/L) | Partition gravimetric method |
| 31. | Hydrocarbons (ppb) (μg/L) | Fluorescence spectrophotometer |
| 32. | Pesticides (ppm) | Liquid-liquid extraction gas chromatography |
| 33. | Fluoride (ppm) | SPANDS method/Ion selective electrode method |
| 34. | Total organic carbon of water (mg/L) | Combustion-infrared method |
| 35. | Phenol | Chloroform extraction method |
| 36. | Zooplankton | Hanging drop method/Sedgwick Rafter cell |
| 37. | Phytoplankton | Lacky's drop method/Sedgwick Rafter cell |
| 38. | Cyanide | Cyanide selective electrode |
| 39. | Total coliform (CFU/100 mL) | Membrane filter method |
| 40. | Faecal coliform (CFU/100 mL) | Membrane filter method |

## Appendix IX

Table 14. Methods and detection limit for analysis of water parameters

| Parameter | Methodology | Detection limit |
|---|---|---|
| pH | Electrometric method | 0.1 |
| Conductivity | Conductometric method | Less than 10 μS/cm |
| Turbidity | Nephelometric method | Less than 1 NTU (about 0.02 NTU) |
| Total alkalinity | Titration method | Less than 20 mg/L as of $CaCO_3$ |
| Total dissolved solids | Gravimetric method | — |
| Total suspended solids | Gravimetric method | — |
| Chlorides | Argentometric method | — |
| Sulphate | Turbidimetric method | 1 mg $SO_4^{2-}$/L |
| Hardness | EDTA titrimetric method (Complexometric titration) | — |
| Sodium | Flame photometric method | 5 μg/L |
| Potassium | Flame photometric method | 0.1 mg/L |
| Nitrate | UV Visible spectrophotometer screening method | — |
| Phosphate as Total phosphorous | Stannous chloride method (Spectrophotometrically) | 3 μg P/L |
| Dissolved oxygen | Winkler method with azide modification | — |
| Chemical oxygen demand | Closed reflux method | — |
| Biochemical oxygen demand | 5 days BOD at 20°C | 2 mg/L for a minimum DO depletion of 2 mg/L |

# Appendix X

**Table 15.** Heavy metals detection limits in water matrix

| Heavy metal | Detection limit | |
|---|---|---|
| | AAS | ICP |
| Cadmium | 0.004-1.8 ppm | 1-5 ppb |
| Chromium | 0.03-10 ppm | 1-5 ppb |
| Copper | 0.01-4 ppm | 1-5 ppb |
| Lead | 0.1-12 ppm | 11-25 ppb |
| Iron | 0.05-8 ppm | 1-5 ppb |
| Manganese | 0.015-4 ppm | < 1ppb |
| Zinc | 0.005-2 ppm | 1-5 ppb |

AAS : Atomic Absorption Spectrophotometer, ICP : Inductively Coupled Plasma.

**Source**: Standard methods for the examination of water and wastewater (20th Edition) (APHA) American Public Health Association (AWWA) American Water Works Association.

## Appendix XI

**Table 16.** Summary of water quality parameters

| Parameter | Indicator | Type of titration | Titrant used | Standardization with | Colour before titration | Colour after titration |
|---|---|---|---|---|---|---|
| Dissolved Oxygen | Starch | Iodometric | Std. Sodium thiosulphate (Hypo solution) | Potassium dichromate | Blue | Colourless |
| Acidity | Methyl orange and phenolphthalein | Acid base titration/ potentiometric titration with indicator | Sodium hydroxide (0.02 *N*) (strong base) | Potassium hydrogen phthalate (Weak acid) | | |
| Alkalinity | Methyl orange and phenolphthalein | Acid base titration/ potentiometric titration with indicator | Sulphuric acid (0.02 *N*) (Strong acid) | Sodium carbonate (weak base) | Yellow | Orange |
| Chloride | Potassium chromate | Precipitation titration/ Argentiometric titration | Silver nitrate (0.0141 *N*) | Sodium chloride (0.0141 *N*) | Light yellow | Brick red (precipitate form) |
| Fluoride | — | Ion selective electrode/SPANDS method | — | — | — | — |
| Sulphate | Barium chloride (Crystals) | Precipitation method by spectrophotometric method at 420 nm | — | — | — | — |

| Parameter | Indicator | Type of titration | Titrant used | Standardization with | Colour before titration | Colour after titration |
|---|---|---|---|---|---|---|
| Sulphide | Starch | Iodometric titration | Sodium thiosulphate | Potassium chromate | — | — |
| Hardness (Total) | EBT | Complexometric titration | EDTA (ethylene diamine tetra acetic acid) | Standard calcium solution | Wine red | Blue |
| Hardness (Calcium) | Murexide | Complexometric titration | EDTA (ethylene diamine tetra acetic acid) | Standard calcium solution | Pink | Purple |
| Chemical oxygen demand | Ferrion | Redox titration and back titration | Ferrous ammonium sulphate | | Wine red | Green |
| Biochemical oxygen demand | Starch | Idometric titration | Std. Sodium thiosulphate (Hypo solution) | Potassium dichromate | Blue | Colourless |
| Total nitrogen | Digestion reagent | Kjeldahl distillation | Sulphuric acid (0.02 N) | Sodium carbonate | Green | Violet |
| Phosphate | Phenolphthalein and Stannous chloride indicator | Acid digestion then neutralization followed by stannous chloride at 690 nm | 6 N sodium hydroxide and ammonium molybdate and stannous chloride (one drop) | | Pink colour after addition of ammonium molybdate colourless | Blue colour |
| Nitrogen as nitrate | No indicator, spectrophotometer absorbance at 220 and 275 nm | UV Visible | 1 mL 1 *N* HCl | Take a 25 mL of sample + 1 *N* 1 mL HCl absorbance at 275 nm | — | — |

## *Appendix XII*

**Table 17.** Indicator used in acidity and alkalinity titration

| Indicator | Colour change | | pH range |
|---|---|---|---|
| | Acid | Alkaline | |
| Methyl orange | Red | Yellow | 3.1-4.4 |
| Phenolphthalein | Colourless | Pink | 8.3-10 |

**Table 18.** Water classification on the basis of hardness

| Water type | Level of hardness | Equivalent $CaCO_3$ |
|---|---|---|
| Soft | Less than 1 | <50 mg/L |
| Moderate hardness | 1-3 | 50-150 mg/L |
| Hard water | 3-6 | 150-300 mg/L |
| Very hard water | Over 6 | >300 mg/L |

**Table 19.** Conductivity conversion

| Multiply ➡ | To get |
|---|---|
| to get ⬅ | Divide |
| µS/cm | 1 µmho/cm |
| mS/cm | 1000 µS/cm |
| µS/cm | 0.5 ppm |

# Appendix XIII

**Table 20.** Methods for soil analysis

| No. | Parameter | Method of analysis |
|---|---|---|
| **Physical parameter** | | |
| 1 | Particle size (Texture) | International pipette method |
| 2 | Bulk density | Core or clod method |
| 3 | Aggregate size distribution | Wet sieving Yoder method |
| 4 | Permeability | Constant head method |
| **Chemical parameter** | | |
| 5 | pH of saturated soil paste | pH measurement |
| 6 | Electrical conductivity of saturated extract | Electrical conductivity meter |
| 7 | Calcium and magnesium | EDTA complexometric titration |
| 8 | Sodium and potassium | Flame photometer |
| 9 | Carbonate ($CO_3$) and bicarbonate ($HCO_3$) | Acid base titration method |
| 10 | Chloride ($Cl^-$) | Mohor's method |
| 11 | Cation exchange capacity (CEC) | • Ammonium acetate method<br>• Sodium acetate method<br>• $BaCl_2$-triethanolamine method |
| 12 | Available nitrogen (N) | • Soil extraction with KCl and estimation of $NH_4$-N and $NO_3$-N by distillation<br>• Specific ion electrode method |
| 13 | Available phosphate | Olson's method |
| 14 | Organic carbon (%) and organic matter (%) | Walkley and Black method |

**Table 21.** Electrical conductivity (EC) of soil and plant growth

| EC range | Rating |
|---|---|
| 0-1 mS/cm | Good soil |
| 1-2 mS/cm | Poor seed emergence |
| 2-3 mS/cm | Harmful to some crops |
| >3 mS/cm | Harmful to most of the crops |

## Appendix XIV

**Table 22.** Soil rating on the basis of organic carbon

| Soil organic carbon | Soil status |
|---|---|
| Less than 0.20% | Very low |
| 0.21 to 0.40% | Low |
| 0.41 to 0.60% | Moderate |
| 0.61 to 0.80% | Moderately high |
| 0.81 to 1.00% | High |
| >1.00% | Very high |

**Table 23.** Calorific values of different fuels

| Fuel | Calorific value |
|---|---|
| Indian coal | Approximately 4500 kcal/kg or 4390 kcal/kg (sulphur content 0.5-0.7% by weight) |
| Heavy furnace oil (HFO) | 10,160 kcal/kg |
| LHSH | 9500 kcal/kg |
| LDO | 10,000 kcal/kg |
| HSD | ~10,000 kcal/kg |

**To convert measured gas concentration from mg/m$^3$ to ppm (at T °C and P atm.)**

Measured concentration (mg/m$^3$) = Calculated concentration (in ppm) x molecular weight/molar volume

Molar volume at 25°C is 24.45 dm$^3$.

Hence,

$SO_2$ = 64.0628, 1 ppm = 2.620 mg/Nm$^3$ = 2620 µg/m$^3$ at 25°C

$H_2S$ = 46.0055, 1 ppm = 1.882 mg/Nm$^3$ = 1882 µg/m$^3$ at 25°C

$O_3$ = 47.9982, 1 ppm = 1.963 mg/Nm$^3$ = 1963 µg/m$^3$ at 25°C

$NH_3$ = 17.00, 1 ppm = 0.6953 mg/Nm$^3$ = 695.3 µg/m$^3$ at 25°C

CO = 28.00, 1 ppm = 1.1452 mg/Nm$^3$ = 1145.2 µg/m$^3$ at 25°C

Conversion of percentage to ppm

100% = 1,000,000 ppm

1% = 10,000 ppm

0.1% = 1,000 ppm

0.01% = 100 ppm

0.001% = 10 ppm

# Appendix XV

## National Ambient Air Quality Standards (Revised)

(MoEF Notification G.S.R 826(E), dated 16.11.2009)

| S. No. | Pollutant | Time Weighted Average | New Standards Schedule VII Rule 3 (3B), 16 Nov. 2009 — Concentration in ambient air: Industrial area, Residential, Rural & other Areas | Ecologically sensitive area (notified by Central Government) | Methods of measurement |
|---|---|---|---|---|---|
| 1 | Sulphur Dioxide ($SO_2$) | Annual Avg* | 50.0 μg/m$^3$ | 20.0 μg/m$^3$ | – Improved West and Geake method |
| | | 24 hours** | 80.0 μg/m$^3$ | 80.0 μg/m$^3$ | – Ultraviolet fluorescence |
| 2 | Oxides of Nitrogen as $NO_2$ | Annual Avg* | 40.0 μg/m$^3$ | 30.0 μg/m$^3$ | – Modified Jacob and Hochheiser (Sodium Arsenite ) |
| | | 24 hours** | 80.0 μg/m$^3$ | 80.0 μg/m$^3$ | – Chemiluminescence |
| 3 | Particulate matter (size less than 10μm) | Annual Avg* | 60.0 μg/m$^3$ | 60.0 μg/m$^3$ | – Gravimetric |
| | | 24 hours** | 100.0 μg/m$^3$ | 100.0 μg/m$^3$ | – TOEM<br>– Beta attenuation |
| 4 | Particulate matter (size less than 2.5 μm) | Annual Avg* | 40.0 μg/m$^3$ | 40.0 μg/m$^3$ | – Gravimetric |
| | | 24 hours** | 60.0 μg/m$^3$ | 60.0 μg/m$^3$ | – TOEM<br>– Beta attenuation |
| 5 | Lead (Pb) | Annual Avg* | 0.50 μg/m$^3$ | 0.50 μg/m$^3$ | – AAS/ICP method for sampling on EPM2000 or Equivalent Filter paper |
| | | 24 hours** | 1.0 μg/m$^3$ | 1.0 μg/m$^3$ | – ED-XRF using Teflon filter paper |

| S. No. | Pollutant | Time Weighted Average | New Standards Schedule VII Rule 3 (3B), 16 Nov. 2009 — Concentration in ambient air: Industrial area, Residential, Rural & other Areas | Concentration in ambient air: Ecologically sensitive area (notified by Central Government) | Methods of measurement |
|---|---|---|---|---|---|
| 6 | Carbon monoxide (CO) | 8 hours** | 2.0 mg/m$^3$ | 2.0 mg/m$^3$ | – Non Dispersive Infra Red (NDIR) spectroscopy |
| | | 1 hour | 4.0 mg/m$^3$ | 4.0 mg/m$^3$ | |
| 7 | Ozone | 8 hours** | 100.0 µg/m$^3$ | 100.0 µg/m$^3$ | – Photometric |
| | | 1 hour | 180.0 µg/m$^3$ | 180.0 µg/m$^3$ | – Chemiluminescence |
| | | 24 hours** | 60.0 µg/m$^3$ | 60.0 µg/m$^3$ | – Chemical method |
| 8 | Ammonia ($NH_3$) | Annual Avg* | 100.0 µg/m$^3$ | 100.0 µg/m$^3$ | – Chemiluminescence |
| | | 24 hours** | 400.0 µg/m$^3$ | 400.0 µg/m$^3$ | – Indo-Phenol Blue method |
| 9 | Benzene | Annual Avg* | 5.0 µg/m$^3$ | 5.0 µg/m$^3$ | – GC based continuous analyzer |
| | | | | | – Adsorption/desorption followed by GC analysis |
| 10 | Benzo(a) pyrene | Annual Avg* | 1.0 ng/m$^3$ | 1.0 ng/m$^3$ | – Solvent extraction followed by GC/HPLC extraction |
| 11 | Arsenic | Annual Avg* | 6.0 ng/m$^3$ | 6.0 ng/m$^3$ | – AAS/ICP method for sampling on EPM2000 OR Equivalent Filter paper |
| 12 | Nickel | | 20.0 ng/m$^3$ | 20.0 ng/m$^3$ | – AAS/ICP method for sampling on EPM2000 OR Equivalent Filter paper |

* Annual Arithmetic mean of minimum 104 measurements in a year taken twice a week 24 hourly at uniform interval,

** 24 hourly / 8 hourly or 1 hourly monitored values as applicable shall be complied with 98% of the time in a year. However, 2% of the time, they may exceed the limits but not on two consecutive days of monitoring.

## *Appendix XVI*

### Diesel Generator Sets: Stack Height

The minimum height of stack to be provided with each generator set can be worked out by using the following formula :

$$H = h + 0.2 \times \sqrt{KVA}$$

H = Total height of stack in metre

h = Height of the building in meters where the generator set is installed

KVA = Total generator capacity of the set in KVA.

Based on the above formula the minimum stack height to be provided with different range of generator sets may be categorized as follows :

| For Generator Sets | Total Height of Stack in Meter |
|---|---|
| 50 KVA | Ht. of the building + 1.5 meter |
| 50 – 100 KVA | Ht. of the building + 2.0 meter |
| 100 – 150 KVA | Ht. of the building + 2.5 meter |
| 150 – 200 KVA | Ht. of the building + 3.0 meter |
| 200 - 250 KVA | Ht. of the building + 3.5 meter |
| 250 – 300 KVA | Ht. of the building + 3.5 meter |

Similarly for higher KVA ratings a stack height can be worked out using the above formula.

Evolved by CPCB

[Emission Regulations Part IV : COINDS/26/1986-87]

## Appendix XVII

**Table 24.** A comparative analysis of Indian Standards for Drinking Water Specifications (BIS IS 10500)

| No. | Substance or characteristics | Requirement (Desirable limit) | | | | Permissible limit in the absence of alternative source | | | |
|---|---|---|---|---|---|---|---|---|---|
| | | **1983** | **1991** | **2004** | **2012** | **1983** | **1991** | **2004** | **2012** |
| | **Essential characteristics** | | | | | | | | |
| i) | Colour | 10 | 5 | 5 | 5 | 50 | 25 | 25 | 15 |
| ii) | Odour | Unobjectionable | Unobjectionable | Agreeable | Agreeable | — | — | Agreeable | Agreeable |
| iii) | Taste | Agreeable | Agreeable | Agreeable | Agreeable | — | — | Agreeable | Agreeable |
| iv) | Turbidity | 10 | 5 | 5 | 1 | 25 | 10 | 10 | 5 |
| v) | pH | 6.5 to 8.5 | 6.5 to 8.5 | 6.5 to 8.5 | 6.5 to 8.5 | Upto 9.2 | No relaxation | No relaxation | No relaxation |
| vi) | Total hardness | 300 | 300 | 300 | 200 | 600 | 600 | 600 | 600 |
| vii) | Iron | 0.3 | 0.3 | 0.3 | 0.3 | 1.0 | 1.0 | 1.0 | No relaxation |
| viii) | Chlorides | 250 | 250 | 250 | 250 | 1000 | 1000 | 1000 | 1000 |
| ix) | Residual free chlorine | 0.2 | 0.2 | 0.2 | 0.2 | 0.5 | — | 1.0 | 1.0 |
| x) | Fluoride | | 1.0 | 1.0 | 1.0 | 1.5 | 1.5 | 1.5 | 1.5 |

| No. | Substance or characteristics | Requirement (Desirable limit) | | | | Permissible limit in the absence of alternative source | | | |
|---|---|---|---|---|---|---|---|---|---|
| | | 1983 | 1991 | 2004 | 2012 | 1983 | 1991 | 2004 | 2012 |
| | **Desirable characteristics** | | | | | | | | |
| xi) | Dissolved solids | No standard | 500 | 500 | 500 | No standard | 2000 | 2000 | 2000 |
| xii) | Calcium | 75 | 75 | 75 | 75 | 200 | 200 | 200 | 200 |
| xiii) | Magnesium | 30 | 30 | 30 | 30 | 100 | 100 | 100 | 100 |
| xiv) | Copper | 0.05 | 0.05 | 0.05 | 0.05 | 1.5 | 1.5 | 1.5 | 1.5 |
| xv) | Manganese | 0.1 | 0.1 | | 0.1 | 0.5 | 0.3 | | 0.3 |
| xvi) | Sulphate | 150 | 200 | 200 | 200 | 400 | 400 | 400 | 400 |
| xvii) | Nitrate | 45 | 45 | 45 | 45 | No relaxation | 100 | No relaxation | No relaxation |
| xviii) | Phenolic compounds | No standard | 0.001 | 0.001 | 0.001 | No standard | 0.002 | 0.002 | 0.002 |
| xix) | Mercury | 0.001 | 0.001 | 0.001 | | No relaxation | No relaxation | No relaxation | |
| xx) | Cadmium | 0.01 | 0.01 | 0.003 | 0.003 | No relaxation | No relaxation | No relaxation | No relaxation |
| xxi) | Selenium | 0.01 | 0.01 | 0.01 | 0.01 | No relaxation | No relaxation | No relaxation | No relaxation |

| No. | Substance or characteristics | Requirement (Desirable limit) | | | | Permissible limit in the absence of alternative source | | | |
|---|---|---|---|---|---|---|---|---|---|
| | | 1983 | 1991 | 2004 | 2012 | 1983 | 1991 | 2004 | 2012 |
| xxii) | Arsenic | 0.05 | 0.05 | 0.01 | 0.01 | No relaxation | No relaxation | 0.05 | 0.05 |
| xxiii) | Cyanide | 0.05 | 0.05 | 0.05 | 0.05 | No relaxation | No relaxation | No relaxation | No relaxation |
| xxiv) | Lead | 0.1 | 0.05 | 0.01 | 0.01 | No relaxation | No relaxation | No relaxation | No relaxation |
| xxv) | Zinc | 5.0 | 5.0 | 5.0 | 5.0 | 10.0 | 15.0 | 15.0 | 15.0 |
| xxvi) | Anionic detergents | 0.2 | 0.2 | 0.2 | 0.2 | 1.0 | 1.0 | 1.0 | 1.0 |
| xxvii) | Chromium | 0.05 | 0.05 | | 0.05 | No relaxation | No relaxation | | No relaxation |
| xxviii) | Ploy nuclear aromatic hydrocarbons | — | — | | | — | — | | |
| xxix) | Mineral oil | 0.01 | 0.01 | 0.01 | 0.5 | 0.03 | 0.03 | 0.03 | No relaxation |
| xxx) | Pesticides | Absent | Absent | Separate information on each pesticide is given | Separate information on each pesticide is given | — | 0.001 | Separate information on each pesticide is given | Separate information on each pesticide is given |

| No. | Substance or characteristics | Requirement (Desirable limit) | | | | Permissible limit in the absence of alternative source | | | |
|---|---|---|---|---|---|---|---|---|---|
| | | 1983 | 1991 | 2004 | 2012 | 1983 | 1991 | 2004 | 2012 |
| xxxi) | Radioactive materials | | | | | | | | |
| | a) Alpha emitters | 10-8 | — | 0.1 | 0.1 | — | 0.1 | 0.1 | No relaxation |
| | b) Beta emitters | 10-7 | — | 1.0 | 1.0 | — | 1.0 | 1.0 | No relaxation |
| xxxii) | Alkalinity | No standard | 200 | 200 | 200 | No standard | 600 | 600 | 600 |
| xxxiii) | Aluminium | No standard | 0.03 | 0.03 | 0.03 | No standard | 0.2 | 0.2 | 0.2 |
| xxxiv) | Boron | No standard | 1 | 0.3 | 0.5 | No standard | 5 | 1.5 | 1.0 |
| xxxv) | Barium | No standard | | 0.7 | 0.7 | No standard | | No relaxation | No relaxation |
| xxxvi) | Ammonia | No standard | | | 0.5 | No standard | | | No relaxation |
| xxxvii) | Chloramines | No standard | | | 4.0 | No standard | | | No relaxation |

## Appendix XVIII

### ICMR Standards of Bacteriological Quality of Water*

### Treated Water

Efficient purification of water supply with chlorination should render the supply free from coliform bacteria.

Throughout any month, in 90% of samples examined, coliform bacteria should be absent in 100 mL or the MPN of coliform should be less than 1.0 per 100 mL. No sample should show MPN of coliform in excess of 10 per 100 mL.

MPN of coliform of 8 to 10 per 100 mL should not occur in consecutive samples. If so, additional samples should collected from the same point every day until two consecutive samples show MPN of 8-10 per 100 mL.

If necessary, operation of treatment plant should be checked. Some deterioration in bacterial quality is likely to occur in a water supply when the water has to traverse through an extensive distribution system.

### Untreated Water

Standards of bacterial quality are recommended for untreated water used for human consumption

Throughout any month, none of the samples should show coliform bacteria in excess of 10 per 100 mL.

An MPN of 20 per 100 mL may be permitted occasionally, if *E-coli* is absent

If MPN of coliform bacteria is consistently more than 20, treatment of water supply should be considered

### Rural Supplies

The bacteriological standards of quality recommended for untreated water supply may be applied to the small rural water supplies also.

* Indian Council of Medical Research Special Report Series No. 44 Manual of Standards of Quality for Drinking Water Supplies New Delhi - 1962

# Appendix XIX

## Classification of Inland Surface Water (CPCB Standards)

| | Characteristics | A@ | B@ | C@ | D@ | E@ |
|---|---|---|---|---|---|---|
| 1. | Dissolved oxygen, mg/L, Min | 6 | 5 | 4 | 4 | — |
| 2. | Biochemical oxygen demand, mg/L, Max | 2 | 3 | 3 | | |
| 3. | Total Coliform organisms,* MPN/100 mL, max | 50 | 500 | 5000 | — | — |
| 4. | Total Dissolved Solids, mg/L, Max. | 500 | — | 1500 | — | 2100 |
| 5. | Chlorides (as Cl), mg/L, Max. | 250 | — | 600 | — | 600 |
| 6. | Colour, Hazen units, Max. | 10 | 300 | 300 | — | - |
| 7. | Sodium absorption ratio, Max. | — | — | — | — | 26 |
| 8. | Boron (as B), mg/L, Max. | — | — | — | — | 2 |
| 9. | Sulphates (as $SO_4$), mg/L, Max | 400 | — | 400 | — | 1000 |
| 10. | Nitrates (as $NO_3$), mg/L, Max. | 20 | — | 50 | — | — |
| 11. | Free Ammonia (as N), mg/L, Max. | — | — | — | 1.2 | — |
| 12. | Conductivity at 25 °C, micromhos/cm, Max. | — | — | — | 1.0 | 2.25 |
| 13. | pH value | 6.5-8.5 | 6.5-8.5 | 6.5-8.5 | 6.5-8.5 | 6.0-8.0 |
| 14. | Arsenic (as As), mg/L, Max. | 0.05 | 0.2 | 0.2 | — | — |
| 15. | Iron (as Fe) mg/L, Max. | 0.3 | — | 50.0 | — | — |
| 16. | Fluorides (as F), mg/L, Max. | 1.5 | 1.5 | 1.5 | — | — |
| 17. | Lead (as Pb), mg/L, Max. | 0.1 | — | 0.1 | — | — |
| 18. | Copper (as Cu), mg/L, Max. | 1.5 | — | 1.5 | — | — |
| 19. | Zinc (as Zn), mg/L, Max. | 15.0 | — | 15.0 | — | — |

* If the coliform count is found to be more than the prescribed tolerance limits, the criteria for coliform shall be satisfied if not more than 20 percent of samples show more than the tolerance limits specified, and not more than 5 percent of samples show values more than 4 times the tolerance limits. Further, the faecal coliform should not be more than 20 percent of the coliform. Source: Indian Standard (IS: 2296 - 1982).

@ A - Drinking water source without conventional treatment but after disinfection

B - Outdoor bathing (organised)

C - Drinking water source with conventional treatment followed by disinfection

D - Propagation of Wildlife, Fisheries

E - Irrigation, Industrial cooling, Controlled waste disposal

## *Appendix XX*

### Thermal Power Plant: Standards for Liquid Effluent

| Source | Parameter | Standards Max. limiting concentration, mg/L (except for pH and Temp.) |
|---|---|---|
| Condenser cooling water (once through cooling system) | pH | 6.5 to 8.5 |
| | Temperature | Not more than 5°C higher than the intake |
| | Free available chlorine | 0.5 |
| Boiler blowdown | Suspended solids | 100 |
| | Oil & Grease | 20 |
| | Copper (total) | 1.0 |
| | Iron (total) | 1.0 |
| Cooling tower blowdown | Free available Chlorine | 0.5 |
| | Zinc | 1.0 |
| | Chromium (total) | 0.2 |
| | Phosphate | 5.0 |
| | Other corrosion inhibiting material | Limit to be established on case by case basis by Central Board in case of Union Territories and State Boards in case of States |
| Ash Pond Effluent | pH | 6.5 to 8.5 |
| | Suspended solids | 100 |
| | Oil & Grease | 20 |

*Source:* EPA Notification

[S.O 844(E), dt 19th Nov; 1986]

# Appendix XXI

## General Standards for Discharge of Environment Pollutants: Effluent

(Gazette Notification of MoEF, May 1993)

| S. No. | Parameter | Inland surface water | Public sewers | Land for irrigation | Marine coastal areas |
|---|---|---|---|---|---|
| 1. | Colour and odour | — | — | — | — |
| 2. | Suspended solids (mg/L) | 100 | 600 | 200 | a) For process waste water-100<br>b) For cooling water effluent 10 percent above total suspended matter of influent |
| 3. | Particular size of suspended solids | Shall pass 850 micron IS Sieve | — | | a) Floatable solids, max. 3 mm<br>b) Settable solids, max 850 microns |
| 4. | *** | * | — | *** | — |
| 5. | pH value | 5.5 to 9.0 | 5.5 to 9.0 | 5.5 to 9.0 | 5.5 to 9.0 |
| 6. | Temperature | Shall not exceed 5°C above the receiving water temperature | — | — | Shall not exceed 5°C above the receiving water temperature |
| 7. | Oil and grease mg/L | 10 | 20 | 10 | 20 |
| 8. | Total residual chlorine, mg/L Max. | 1.0 | — | — | 1.0 |
| 9. | Ammoniacal nitrogen | 50 | 50 | — | 50 |
| 10. | Total Kjeldahl Nitrogen (as $NH_3$) mg/L, Max | 100 | — | — | 100 |

| S. No. | Parameter | Inland surface water | Public sewers | Land for irrigation | Marine coastal areas |
|---|---|---|---|---|---|
| 11. | Free ammonia (as $NH_3$) mg/L, Max | 5.0 | — | — | 5.0 |
| 12. | Biochemical Oxygen demand (5 days at 20°C) mg/L Max. | 30 | 350 | 100 | 100 |
| 13. | Chemical Oxygen demand, mg/L Max. | 250 | — | — | 250 |
| 14. | Arsenic (as As), mg/L Max. | 0.2 | 0.2 | 0.2 | 0.2 |
| 15. | Mercury (as Hg) mg/L, Max. | 0.01 | 0.01 | — | 0.01 |
| 16. | Lead (as Pb) mg/L, Max. | 0.1 | 1.0 | — | 2.0 |
| 17. | Cadmium (as Cd) mg/L, Max. | 2.0 | 1.0 | — | 2.0 |
| 18. | Hexavalent chromium (as $Cr^{+}6$), mg/L, Max. | 0.1 | 2.0 | — | 2.0 |
| 19. | Total chromium (as Cr) mg/L, Max. | 2.0 | 2.0 | — | 2.0 |
| 20. | Copper (as Cu) mg/L, Max. | 3.0 | 3.0 | — | 3.0 |
| 21. | Zinc (as Zn.) mg/L, Max. | 5.0 | 15 | — | 15 |
| 22. | Selenium (as Se) mg/L, Max. | 0.05 | 0.05 | — | 0.05 |

| S. No. | Parameter | Inland surface water | Public sewers | Land for irrigation | Marine coastal areas |
|---|---|---|---|---|---|
| 23. | Nickel (as Ni) mg/L, Max. | 3.0 | 3.0 | — | 5.0 |
| [2]24. | *** | * | * | * | * |
| [2]25. | *** | * | * | * | * |
| [2]26. | *** | * | * | * | * |
| 27. | Cyanide (as CN) as mg/L, Max. | 0.2 | 2.0 | 0.2 | 0.2 |
| [2]28. | *** | * | * | * | * |
| 29. | Fluoride (as F) as mg/L, Max. | 2.0 | 15 | — | 15 |
| 30. | Dissolved phosphates (as P), mg/L Max. | 5.0 | — | — | — |
| [2]31. | *** | * | * | * | * |
| 32. | Sulphide (as S) mg/L, Max. | 2.0 | — | — | 5.0 |
| 33. | Phenolic compounds (as $C_6H_5OH$) mg/L Max. | 1.0 | 5.0 | — | 5.0 |
| 34. | Radioactive materials | | | | |
| | (a) Alpha emitter micro curie/mL | $10^{-7}$ | $10^{-7}$ | $10^{-8}$ | $10^{-7}$ |
| | (b) Beta emitter micro curie/mL | $10^{-6}$ | $10^{-6}$ | $10^{-7}$ | $10^{-7}$ |
| 35. | Bio-assay test | 90% survival of fish after 96 hours in 100% effluent | 90% survival of fish after 96 hours in 100% effluent | 90% survival of fish after 96 hours in 100% effluent | 90% survival of fish after 96 hours in 100% effluent |

| S. No. | Parameter | Inland surface water | Public sewers | Land for irrigation | Marine coastal areas |
|---|---|---|---|---|---|
| 36. | Manganese (as Mn) mg/L, Max. | 2 | 2 | — | 2 |
| 37. | Iron (as Fe) mg/L, Max. | 3 | 3 | — | |
| 38. | Vanadium (as V) mg/L, Max. | 0.2 | 0.2 | — | 0.2 |
| 39. | Nitrate Nitrogen mg/L, Max. | 10 | — | — | 20 |
| [2]40. | *** | * | * | * | * |

1. Schedule : VI inserted by Rule 2(d) of the Environment (Protection), Second Amendment Rules, 1993 notified vide G.S.R. 422(E) dated 19.05.1993, published in the Gazette No. 174 dated 19.05.1993.
2. Omitted by Rule 2(d) (i) of the Environment (Protection) Act Third Amendment Rules, 1993 vide Notification No. G.S.R. 801 (E) dated 31.12.1993

## Suitability of Treated Wastewater for Irrigation

| | Tolerance limit for inland surface water for irrigation | |
|---|---|---|
| | MoEF, 1993 | IS 2296, 1982 |
| pH | 5.5-9.0 | 6.0-8.5 |
| Dissolved solids mg/L | — | 2100 |
| Electrical Conductivity *(EC) µS/cm | — | Class-$C_3$ |
| Sodium mg/L | — | — |
| Percent sodium | — | 60 |
| Sodium adsorption ration (*SAR) | — | Class-$S_1$ |
| Sulphates mg/L | — | 1000 |
| Chlorides mg/L | — | 600 |
| Oil and grease mg/L | 10 | — |
| Biochemical oxygen demand (BOD) mg/L | 100 | — |
| Suspended solids mg/L | 200 | — |
| Bioassay | 90% survival for 96 hrs in 100% effluent | — |

| * EC | Class | SAR | Class |
|---|---|---|---|
| 100-200 | $C_1$ | <10 | $S_1$ |
| 250-750 | $C_2$ | 10-18 | $S_2$ |
| 750-2250 | $C_3$ | 18-26 | $S_3$ |
| 2250-5000 | $C_4$ | >26 | $S_4$ |

## Appendix XXII

### National Ambient Air Quality Standards in Respect of Noise (CPCB)

| Area Code | Category of Area/Zone | Limits in dB (A) Leq* | |
|---|---|---|---|
| | | Day Time | Night Time |
| (A) | Industrial Area | 75 | 70 |
| (B) | Commercial Area | 65 | 55 |
| (C) | Residential Area | 55 | 45 |
| (D) | Silence Zone | 50 | 40 |

**Notes :**

1. Day time shall mean from 6.00 a.m. to 10.00 p.m.
2. Night time shall mean from 10.00 p.m. to 6.00 a.m.
3. Silence zone is defined as an area comprising not less than 100 meters around Hospitals, Educational Institutions and courts. The silence zones are zones which are declared as such by the competent authority.
4. Mixed categories of areas may be declared as one of the four abovementioned categories by the Component Authority.

* dB(A) Leq denotes the time weighted average of the level of sound in decibels on scale A which is related to human hearing

"A", in dB(A) Leq, denotes the frequency weighting in the measurement of noise and corresponds to frequency response characteristics of human ear

Leq: It is an energy mean of the noise level over a specified period

(The Principal Rules were published in the Gazette of India, vide S.O. 123(E), dated 14.2.2000 and subsequently amended vide S.O. 1046(E), dated 22.11.2000, S.O. 1088(E), dated 11.10.2002, S.O. 1569 (E), dated 19.09.2006 and S.O. 50 (E) dated 11.01.2010 under the Environment (Protection) Act, 1986).

# Appendix XXIII

## Damage Risk Criteria for Hearing Loss Occupational Safety & Health Administration (OSHA)

| Maximum Allowable Duration Per Day, Hour | Noise Level dB (A) (Slow Response) |
|---|---|
| 8 | 90 |
| 6 | 92 |
| 4 | 95 |
| 3 | 97 |
| 2 | 100 |
| 1.5 | 102 |
| 1 | 105 |
| 0.5 | 110 |
| 0.25 or less | 115 |

## Noise Standards for Fire-crackers

A. (i) The manufacture, sale or use of fire-crackers generating noise level exceeding 125 dB (AI) or 145 dB (C) pk at 4 meters distance from the point of bursting shall be prohibited.

(ii) For individual fire-cracker constituting the series (joined fire-crackers), the above mentioned limit be reduced by 5 log10 (N) dB, where N = number of crackers joined together.

B. The broad requirements for measurement of noise from fire-crackers shall be-

(i) The measurements shall be made on a hard concrete surface of minimum 5 meter diameter or equivalent.

(ii) The measurements shall be made in free field conditions i.e., there shall not be any reflecting surface upto 15 meter distance from the point of bursting.

(iii) The measurement shall be made with an approved sound level meter.

Note:

dB (AI) : A-weighted impulse Sound Pressure Level in decibel

dB (C) pk : C-weighted Peak Sound Pressure Level in decibel.

**Table 25.** Noise limits for vehicles applicable at manufacturing stage from year 2003

| S. No. | Type of vehicle | Noise Limits from 1st January, 2003, dB(A) | Date of implementation |
|---|---|---|---|
| (1) | (2) | (3) | (4) |
| 1. | Two wheeler | | 1st January, 2003 |
| | Displacement upto 80 $cm^3$ | 75 | |
| | Displacement more than 80 $cm^3$ but upto 175 $cm^3$ | 77 | |
| | Displacement more than 175 $cm^3$ | 80 | |
| 2. | Three wheeler | | 1st January, 2003 |
| | Displacement upto 175 $cm^3$ | 77 | |
| | Displacement more than 175 $cm^3$ | 80 | |
| 3. | Passenger car | 75 | 1st January, 2003 |
| 4. | Passenger or commercial vehicle | | 1st July, 2003 |
| | Gross vehicle weight upto 4 tonne | 80 | |
| | Gross vehicle weight more than 4 tonne but upto 12 tonne | 83 | |
| | Gross vehicle weight more than 12 tonne | 85 | |

**Table 26.** Noise limits for vehicles applicable at manufacturing stage applicable from 1st April, 2005

| S. No. | Type of vehicle | Noise Limits from 1st January, 2003, dB(A) |
|---|---|---|
| 1.0 | Two wheeler | |
| 1.1 | Displacement upto 80 cc | 75 |
| 1.2 | Displacement more than 80 cc but upto 175 cc | 77 |
| 1.3 | Displacement more than 175 cc | 80 |
| 2.0 | Three wheeler | |
| 2.1 | Displacement upto 175 cc | 77 |
| 2.2 | Displacement more than 175 cc | 80 |
| 3.0 | Vehicles used for carriage of passengers and capable of having not more than nine seats, including the driver's seat | 74 |
| 4.0 | Vehicles used for carriage of passengers having more than nine seats, including the driver's seat, and a maximum gross Vehicle Weight(GVW) of more than 3.5 tonnes | |
| 4.1 | With an engine power less than150 KW | 78 |
| 4.2 | With an engine power of 150 KW or above | 80 |
| 5.0 | Vehicles used for carriage of passengers having more than nine seats, including the driver's seat: Vehicles used for carriage goods. | |
| 5.1 | With maximum GVW not exceeding 2 tonnes | 76 |
| 5.2 | With maximum GVW greater than 3 tonnes but not exceeding 3.5 tonnes | 77 |
| 6.0 | Vehicles used for transport of goods with a maximum GVW exceeding 3.5 tonnes | |
| 6.1 | With an engine power less than 75 KW | 77 |
| 6.2 | With an engine power of 75 KW or above but less than 150 KW | 78 |
| 6.3 | With an engine power of 150 KW or above. | 80 |

## Appendix XXIV

### Environment calendar

**Environment related days**

January 5$^{th}$ : National Birds Day

February 2$^{nd}$ : World Wetland Day

March 3$^{rd}$ : World Wildlife Day

Second Friday of March : Solar Appreciation Day

March 14$^{th}$ : International Day of Action for Rivers

March 20$^{th}$ : World Sparrow Day

March 21$^{st}$ : World Forest Day

March 21$^{st}$ : World Planting Day

March 21$^{st}$ : World Wood Day

March 22$^{nd}$ : World Water Day

March 23$^{rd}$ : World Meteorological Day

March (towards end of the month) : Earth Hour Day (8.30 pm to 9.30 pm)

April 7$^{th}$ : World Health Day

April 18$^{th}$ : World Heritage Day

April 22$^{nd}$ : Earth Day

Second Saturday in May : World Migratory Bird Day

Third Friday in May : Endangered Species Day

Third Friday in May : Bike-to-Work Day

May 17$^{th}$ : International Recycle Day

May 22$^{nd}$ : World Biodiversity Day

June 5$^{th}$ : World Environment Day

June 8$^{th}$ : World Oceans Day

June 9$^{th}$ : Coral Triangle Day

June 15$^{th}$ : Global Wind Day

June 16$^{th}$ : World Sea Turtle Day

June 17$^{th}$ : World Day to Combat Desertification and Drought

June 21$^{st}$ : World Giraffe Day

July 1$^{st}$ week : *Vanmohatsav*

July 11$^{th}$ : World Population Day

July 28$^{th}$ : World Nature Conservation Day

July 29$^{th}$ : International Tiger Day

August 6th : Hiroshima Day
August 9th : Nagasaki Day
August 10th : World Lion Day
August 12th : World Elephant Day
August 19th : World Orangutan Day
August 22nd : National Honey Bee Day
First Saturday in September : International Vulture Awareness Day
September 16th : World Ozone Day
September 18th : World Water Monitoring Day
September 21st : Zero Emissions Day
September 22nd : Car Free Day
September 22nd : World Rhino Day
September 26th : World Environmental Health Day
September 28th : Green Consumer Day
Last Sunday in September : World River Day
First Monday in October : World Habitat Day
October 2nd-8th : Wildlife Week
October 3rd : World Nature Day
October 4th : World Animal Day
October 6th : World Wildlife Day
Second Wednesday in October : International Day for Natural Disaster Reduction
October 16th : World Food Day
Fourth Wednesday of October : Sustainable Day
October 23rd : International Snow Leopard Day
October 24th : International Day of Climate Action
November 1st : World Vegan Day
November 5th : World Tsunami Awareness Day
November 6th : International Day for Preventing the Exploitation of the Environment in War and Armed Conflict
November 15th : International GIS Day
December 2nd : National Pollution Prevention Day
December 5th : World Soil Day
December 11th : International Mountain Day
December 14th : National Energy Conservation Day

## Environment related years

| Name | Year |
|---|---|
| International Year of Pulses | 2016 |
| International Year of Soils | 2015 |
| International Year of Forests | 2011 |
| International Year of Biodiversity | 2010 |
| Year of the Gorilla | 2009 |
| International Year of Natural Fibres | 2009 |
| International Year of Planet Earth | 2008 |
| International Year of Sanitation | 2008 |
| International Polar Year | 2007-2008 |
| International Year of the Dolphin | 2007-2008 |
| International Polar Year | 2007-2008 |
| International Year of Deserts and Desertification | 2006 |
| International Year of Freshwater (IYF) | 2003 |
| International Year of Ecotourism (IYE) | 2002 |
| International Year of Mountains (IYM) | 2002 |
| International Year of the Ocean (IYO) | 1998 |
| World Population Year | 1974 |
| International Polar Year | 1932–1933 |
| International Polar Year | 1882–1883 |

## Environment related decades

| Name | Decade |
|---|---|
| United Nations Decade on Biodiversity | 2011-2020 |
| United Nations Decade for Deserts and the Fight against Desertification | 2010-2020 |
| Water for Life Decade | 2005-2015 |
| United Nations Decade of Education for Sustainable Development | 2005-2014 |
| International Decade for Natural Disaster Reduction | 1990s |
| International Drinking Water Decade, 1981-1990 | 1980s |

⊙